CW00485568

BRITAIN'S HOVERFLIES

A field guide

Stuart Ball and Roger Morris

WILDGuides

PRINCETON
press.princeton.edu

Published by Princeton University Press,
41 William Street, Princeton, New Jersey 08540
In the United Kingdom: Princeton University Press, 6 Oxford Street,
Woodstock, Oxfordshire OX20 1TW
nathist.press.princeton.edu

Requests for permission to reproduce material from this work should be sent to
Permissions, Princeton University Press

First published 2013
Second Edition 2015

Copyright © 2013, 2015 Stuart Ball and Roger Morris

Copyright in the photographs remains with the individual photographers.

All rights reserved. No part of this publication may be reproduced, stored in a
retrieval system, or transmitted, in any form or by any means, electronic, mechanical,
photocopying, recording, or otherwise, without the prior permission of the publishers.

British Library Cataloging-in-Publication Data is available

Library of Congress Control Number 2014954503
ISBN 978-0-691-16441-0

Production and design by **WILD***Guides* Ltd., Old Basing, Hampshire UK.
Printed in China

10 9 8 7 6 5 4 3 2

Contents

Syrphus on ragwort.

Foreword

Hoverflies have long been regarded as nice insects, charismatic even. And they get a good press as man's friend, not only because they are colourful and tamely sit on flowers, but because many are useful to gardeners in controlling the numbers of aphids attacking plants.

Yet despite this popularity, many people would not be sure whether they are looking at a hoverfly or not. Many species are excellent mimics of bees and wasps, and such is the variety of appearance that many other species would not be recognised as hoverflies at all.

One of the big incentives to looking at hoverflies has been the rapid rise in popularity of close-up digital photography. Many people are discovering a new wonderful world of small creatures which are amazing when seen in close-up detail on a monitor screen. That is step one but the enquiring mind then asks 'but what is this marvellous fly and what does it do for a living?'

This is where the Hoverfly WildGuide becomes invaluable. A set of high quality photos of live insects is designed to facilitate identification, accompanied by information on life history and distribution. The photos also illustrate the angles of view that best show identification features.

Many other books include a small selection of example hoverflies, in some cases wrongly identified or with erroneous drawings. This **WILD***Guides* publication is unique in that it covers all the genera and a high proportion of the British species, with accurately named photos, and is authored by the co-organisers of the Hoverfly Recording Scheme.

There is not space to cover every species recorded in Britain, and indeed on average about one new species is added to the British list each year. Most of the omitted species are either unlikely to be encountered or cannot be accurately identified on the basis of a photo. Full coverage is provided in the mini-monograph *British Hoverflies* published by the British Entomological and Natural History Society.

Hoverflies are one of the most diverse and fascinating groups of insects in their range of life styles. This makes them an especially important group of animals for monitoring the health of the countryside, and indeed town, and for monitoring the biodiversity response to climate change and various other environmental issues. But the main purposes of this WildGuide are to help you to share in the pleasure of looking at hoverflies, and to provide encouragement to contribute to knowledge of our fauna via the Hoverfly Recording Scheme.

Alan Stubbs

Scaeva pyrastri.

Preface

Sales of the first edition of *Britain's Hoverflies* greatly exceeded our expectations. Although it was only published in 2013, by spring 2014 there was a need to consider either reprinting the original or revising the book. We felt that a revision was essential because so much has changed in the short time since the original text was finalised in 2012. In addition, we wanted to provide some new identification tips, correct a few errors and react to feedback from readers that had highlighted instances where additional explanation was needed. To this end, we have expanded the text within some of the species accounts and included a number of new photographs to help resolve areas of confusion.

Some readers have asked why we did not illustrate all the British species. The reason for this is that we intended this guide to be an introduction to hoverflies that would complement rather than compete with British and northern European monographs (*British Hoverflies: An Illustrated Identification Guide* by Alan E. Stubbs & Steven J. Falk, and *Hoverflies of Northwest Europe: Identification Keys to the Syrphidae* by M. P. van Veen). Both these books remain in print and are essential reading for those who wish to identify difficult taxa within genera such as *Cheilosia*, *Platycheirus* and *Pipiza*, for example. Feedback from various specialists in Europe confirms that our aspiration has
been largely met in providing a readily accessible guide at a price that does not put beginners off and introduces them to these fascinating insects.

Hoverfly taxonomy is fast-moving and there have already been revisions to the number of genera. *Arctophila* and *Chamaesyrphus* have been sunk into *Sericomyia* and *Pelecocera* respectively. In addition, there has been a substantial revision of the names within *Pipiza*, although those for the species illustrated have not changed. By the end of 2013, two further species had been added to the British list (*Scaeva dignota* and *Eumerus sogdianus*), both of which had been expected for some while.

Whilst making these revisions, we also felt that there was scope to expand some aspects of the coverage of the text and illustrations. Over the past few years we have extracted records from over 25,000 photographs posted on various websites, and from these have been able to assess which species are most frequently photographed. This has led to the preparation of a set of new plates covering these species, which we hope will help photographers develop their identification skills. We have also taken the opportunity to include a new section on hoverfly photography, since it is clear from the large number of pictures posted on websites that this is an area of growing interest. In addition, we have added a brief overview of some aspects of the ways in which the data compiled by the Hoverfly Recording Scheme can be used. Users of *Britain's Hoverflies* have an important contribution to make and are encouraged to continue to do so.

Stuart Ball & Roger Morris

December 2014

Introduction

On any fine day between April and October, whether you are in the countryside or in an urban or suburban park or garden, you are likely to come across brightly coloured, black-and-yellow flies hovering around flowers. Although they are trying to convince you they are wasps or bees, they are actually hoverflies. They are such constant flower visitors that, in some other parts of the world, they are called 'flower flies'.

At the time of writing, 283 species of hoverfly have been found in the British Isles, although more are being discovered at an average of about one species per year. Many are quite easily recognisable because of their bright colour patterns. This book aims to introduce this fascinating family by making it as easy as possible to identify some of the commoner and more distinctive species. However, it is important to be aware that a fair proportion of British hoverflies are **not** easy to identify in the field. In many cases, detailed examination of microscopic characters is needed to be certain of their identity. In total, 167 species are illustrated and described in this book, concentrating on the ones you are most likely to find. However, in order to show the full variety of hoverflies, at least one example from each of the 68 genera occurring in Britain is included. The sections on **Identifying Hoverflies** on *page 49* and **Further reading** on *page 302* provide information on where to go next if you want to take things further and tackle some of the more challenging species.

It is also important to understand the way in which hoverflies are classified in order to be able to appreciate fully the relationship and similarities between species. The **Guide to the tribes** on *page 55* attempts to shed some light on this complex and, at first glance, confusing situation and provides a starting point on the path to attempted identification.

This book is illustrated with a combination of field photographs and close-up digital images which show in detail the characters used in separating the species.

Marmalade Fly *Episyrphus balteatus* on Common Ragwort.

Up-to-date distribution maps and diagrams showing the flight periods are derived from the Hoverfly Recording Scheme (see *page 297*). Icons have been included as an indication of the difficulty of identification. Comments in the descriptions are also included to make it clear whether identification is possible in the field.

The availability and sophistication of digital cameras is leading more and more people to take pictures of insects in the field and subsequently seeking to identify them. There is a growing trend for this to be regarded as an alternative to the collection of specimens, but photography remains just one part of a bigger process of making an accurate identification; the species accounts indicate where photographs can be reliably used for this purpose. The arguments around the ethics of collecting specimens are covered on *page 291*.

Apart from being attractive and interesting, hoverflies also play important roles in the environment. Most gardeners know that hoverfly larvae are voracious predators of aphids and are, therefore, 'friends'. However, there are two species (the Greater and Lesser Bulb Flies) whose larvae tunnel in daffodil and other bulbs and are therefore not so welcome in the garden.

Most hoverflies do not have common names. There are a few that do; for example the Drone Fly *Eristalis tenax* and the Marmalade Fly *Episyrphus balteatus*. It has been suggested that common names should be invented for the rest, but this is unlikely to work. Common names catch on because they sum up some aspect of the species' appearance, behaviour or habitat in a way that is memorable. Contrived names seldom manage this and often end up being no more memorable than the scientific name they try to replace. So, the few common names that have become established are used where appropriate but otherwise scientific names have been used.

Like many insects, hoverflies are sensitive indicators of the health of our environment. They are short-lived, fast-breeding and show rapid changes in both range and abundance in response to change: these responses can be detected and interpreted. Some of the changes that have been revealed by analysis of records collected by the Hoverfly Recording Scheme are discussed in later sections of this introduction and in the species accounts.

Hoverflies are the most attractive and accessible group of flies, so you might expect that there would be plenty of popular literature about them. Unfortunately, this is not the case. The standard work is *British Hoverflies* by Alan Stubbs and Steven Falk, originally published in 1983 and fully revised in 2002. However, this is available only through specialist suppliers of entomological works. *A Naturalist's Handbook (No 5)*, by Francis Gilbert in 1993, is not intended as an identification guide but provides much interesting information about their natural history. However, *The Natural History of Hoverflies* by Graham Rotheray and Francis Gilbert is a recent account of what is known about this family apart from identification. It is hoped that this **WILD***Guides* publication will plug this gap and provide an introduction that allows people with at least a passing interest to get started.

Inevitably, a number of technical terms are used when referring to hoverflies. Whilst these are explained in the text where appropriate, for ease of reference all are defined in the **Glossary** starting on *page 46*.

Is it a Hoverfly?

Class	INSECTA			
Order *There are 30 or so Orders in the Class Insecta, of which **Diptera** and **Hymenoptera** are two*	**Diptera** True flies *Within Diptera are around 165 extant families – 107 in the British Isles (of which the **Syrphidae** is one).* **Two (one pair) wings**: See following text	**Hymenoptera** Sawflies, bees, wasps and ants *The Hymenoptera are important in the scope of this book as a number of British hoverfly species mimic species within this order.* **Four (two pairs) wings**: See details below		
Family	**Syrphidae** Hoverflies **vena spuria present**: See following text	**Apidae** Bees (including honey-bees, carpenter bees, cuckoo bees and bumblebees)	**Andrenidae** Solitary bees	**Vespidae** Wasps
	There are around 6,000 species of hoverfly worldwide in 200 genera, of which currently 283 species of 68 genera have been recorded in Britain. This book covers all 68 genera and 167 species in detail.	*In Britain there are ten species of hoverfly that can be regarded as bumblebee mimics, and five or six that can be regarded as honey-bee mimics. Many of the other hoverflies mimic solitary or social wasps and bees in a fairly general way.*		

Hoverfly classification.

Hoverflies are 'true flies', belonging to the very large insect order **Diptera**. The first question must therefore be "Is it a fly?" The clue to recognising true flies is in the name 'Diptera'. It comes from the Greek, 'di' = two and 'ptera' = wings. The characteristic feature of true flies is that they only have two wings (*i.e.* one pair), as compared with most other insects, including bees and wasps, which have four wings (*i.e.* two pairs). In Diptera, the 'hind pair' of 'wings' have evolved into small, club-shaped appendages called '**halteres**' that help control flight and contribute to these insects being such accomplished fliers.

A haltere is visible in this image of the hoverfly *Syritta pipiens*.

A true fly also has the following features:

- A body composed of three sections (like all insects): the head, thorax and abdomen.
- The head has a mouth on the underside, a pair of large, compound eyes, a pair of antennae ('feelers') and usually three 'simple eyes' or ocelli placed at the top, between the compound eyes.
- The thorax bears three pairs of legs and just one pair of wings with the halteres located on each side, just behind and below the wing bases.
- An abdomen, consisting of a series of overlapping segments, with the genitalia located at the tip.

Hoverflies form the family **Syrphidae** and, as with all flies, wing venation is often important in their identification. The veins are rod-like struts that support the flexible membrane that forms the wing surface. Most of the veins run out from the wing base towards the wing tip, but there are also **cross-veins** that form braces between them. Hoverflies have an additional 'spurious vein' (**vena spuria**) which cuts across the normal venation and is actually the reinforced hinge of a longitudinal fold in the wing surface. **The presence of the vena spuria immediately tells you that it is a hoverfly.** This feature is more obvious in some species than others, but there is only one British hoverfly (the rare *Psilota anthracina*) that lacks it altogether.

Another feature of the wing venation is the presence of two outer cross-veins, both near to and parallel with the wing margin, which form a '**false margin**'. Few other Diptera have two outer cross-veins (many have just one or none at all), but the other families with this feature are generally bristly flies.

There are a number of other features that can also help to separate hoverflies from other Diptera:

- They are of fairly compact build, without very elongated bodies, very long legs, a long proboscis or extravagantly developed genitalia.
- Many have black-and-yellow, wasp-like markings or are honey-bee or bumblebee-like, although there are also quite a number of black species. Whilst a few of the darker species have bronzy, greenish or even purplish-metallic reflections, the British species are never brightly metallic blue or green like a green-bottle or blue-bottle fly.
- They may be furry (like a bumblebee – see Mimicry *page 32*), but they are never bristly like house flies or blue-bottles and they do not have bristly legs or strong bristles on the top of the thorax like many flies.
- And, however much they might look like wasps or bees, they are harmless and don't bite or sting!

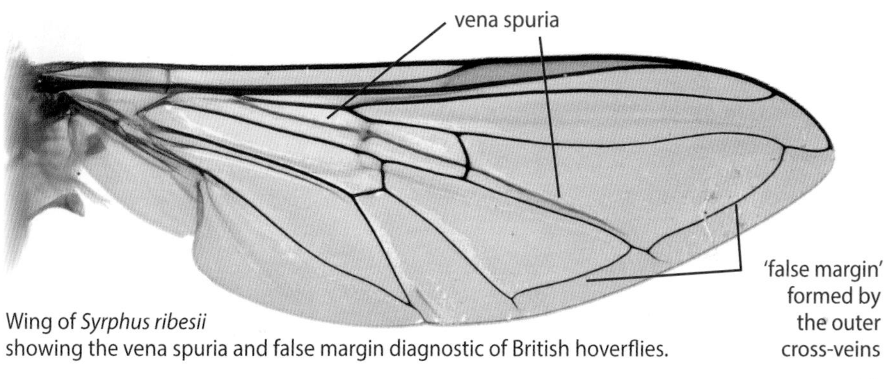

Wing of *Syrphus ribesii*
showing the vena spuria and false margin diagnostic of British hoverflies.

THE LIFE-CYCLE OF A HOVERFLY

Sphaerophoria mating

Episyrphus balteatus laying an egg

Adult

The function of the adult stage is to mate, disperse and lay eggs. Adults may survive from a few days to a few weeks.

Pupa

The pupa is formed inside the skin of the 3rd larval instar and the structure you can see is technically a 'puparium'.

This stage lasts from a few weeks to several months depending on the species.

Egg

Eggs typically hatch a few days after they are laid.

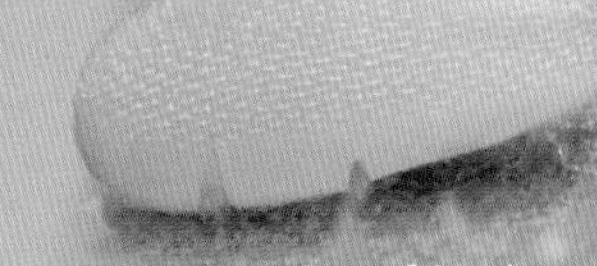
Egg of *Syrphus*

Hoverfly puparium

Larva

The role of the larva is to feed. There are three larval stages or 'instars'. The mature (3rd instar) larva is the stage at which most species spend the majority of the year and during which most of the feeding is done.

Young *Episyrphus* larva amongst aphids

Hoverfly biology

Adults

Great Pied Hoverfly *Volucella pellucens*

The adult stage is what we normally see and think of as 'the hoverfly'. They are relatively short-lived and survive for a few days to a few weeks. One mark-release-recapture exercise revealed that the large, black-and-white Great Pied Hoverfly *Volucella pellucens* can live for at least 35 days, although the average adult life-span was 12 days. Recent mark-recapture studies of *Hammerschmidtia ferruginea* found the maximum adult lifespan to be 55 days. The primary function of the adult is to mate, disperse and lay eggs. Energy and protein are required to form the eggs. In the main, adult hoverflies obtain these by visiting flowers to obtain energy-giving nectar and protein-rich pollen. Most hoverflies lack specialised mouthparts, so prefer to visit flowers in which the nectar and pollen are exposed and easy to reach. White umbellifers, such as Hogweed, and members of the daisy family, such as thistles and knapweeds, are favourites.

One instantly recognisable genus of British hoverflies, *Rhingia*, is exceptional however, in having its mouthparts extended into a long snout, or rostrum. The rostrum has a groove on the underside into which the rather elongated proboscis fits. Consequently, it can visit flowers with a moderately deep tube, like Red Campion and Bluebell, to take advantage of a food source that is out of the reach of other hoverflies.

The closely related genera *Xylota*, *Brachypalpoides* and *Chalcosyrphus* tend not to visit flowers at all. Instead, they feed on honeydew, the sugary secretion of aphids, and pollen grains stuck to the leaf surface. They run over the surface of leaves, rapidly turning backwards and forwards in a very characteristic manner that makes them look like an ichneumon or spider-hunting wasp.

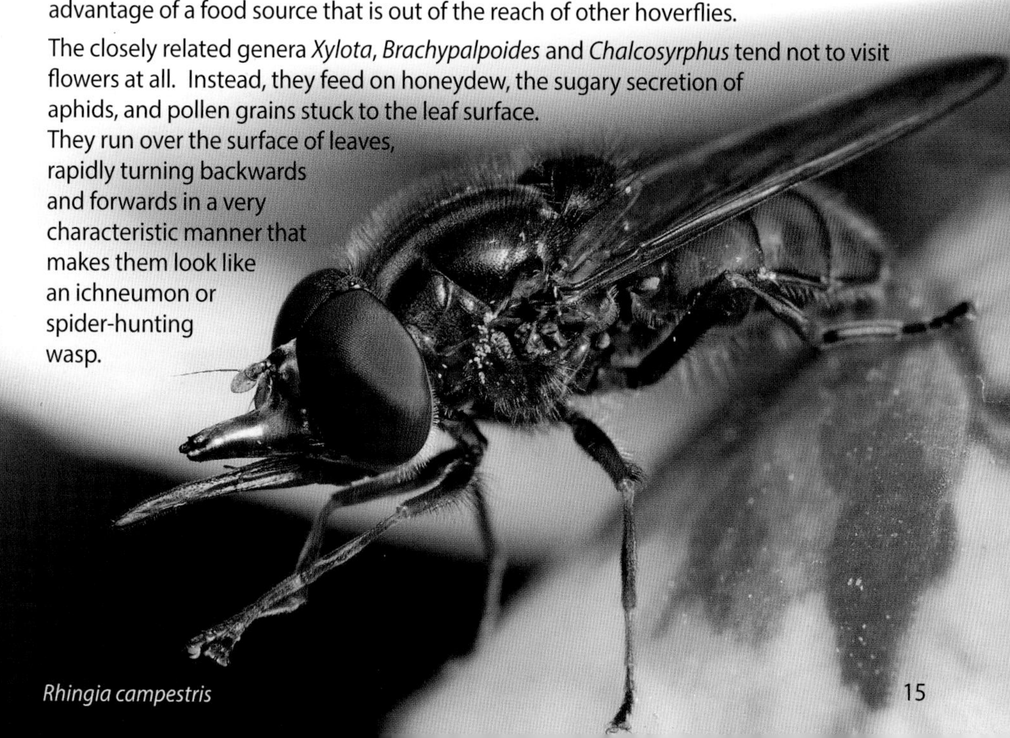
Rhingia campestris

A number of smaller hoverflies, notably members of the tribes Bacchini (*Melanostoma*, *Platycheirus*), visit the flowers of wind pollinated plants such as plantains, grasses and sedges to obtain pollen. These plants provide a particularly rich food source through the production of great quantities of pollen.

Males of many hoverflies show some degree of territorial behaviour. They guard sunny spots by either hovering in a shaft of sunshine or sitting on a sunny leaf, from whence they dart out at other passing insects either to chase off a rival or to mate with a female. The same individual will return repeatedly to the same spot. Mark-release-recapture studies have also found that the same male of *Volucella pellucens* will return to hover in the same position along woodland rides on several successive days, indicating a strong and sophisticated level of territoriality.

Female *Eristalis nemorum* on a knapweed flower with two males hovering above.

The very noticeable high-pitched whine often heard in British woodlands on sunny Summer days is generated by thousands of male *Syrphus* species involved in territorial behaviour. In order to be ready to move at a moment's notice, they keep their wing muscles 'ticking over'. This causes the thorax to vibrate and this vibration is transmitted to the leaf surface, which acts like a sounding board.

Courtship is usually very brief. If the male encounters a female during his darting territorial flights, he grapples with her and, if accepted, they couple, either in flight or after falling onto vegetation. Mating usually lasts for a few minutes.

The small, honey-bee-like hoverfly *Eristalis nemorum* exhibits a more conspicuous courtship-related behaviour than most other hoverflies. A female is often seen feeding on a flower with a male hovering a few inches above her. It is not clear whether this is courtship behaviour or 'mate-guarding'. If a male has mated with a female, it is not unusual, across a range of insect groups, for him to then stand guard over her to stop other males mating with her before she has a chance to lay the eggs he has fertilised. This may be the case with *E. nemorum* but occasionally two or more males may be involved, jockeying for position over the female; this behaviour could therefore be a form of courtship. An interesting project for someone would be to investigate what is really going on!

Occasionally you may encounter a hoverfly with a distended abdomen with distinct pale whitish bands emerging between the segments. Such flies are usually dead and are firmly fixed to a plant; the genera *Melanostoma* and *Platycheirus* are particularly afflicted.

These have succumbed to a fungus that grows inside the insect and emerges between the body plates in order to release its spores. Mass occurrences of dead hoverflies attached to the undersides of umbellifer flowers or to grass stems are most frequently encountered in late Summer and early Autumn. Fungi that attack insects are referred to as 'entomophagous' and are one of the more noticeable causes of death amongst hoverflies.

Melanostoma killed by the fungus *Entomophthora.*

Eggs

The ovipositor of female hoverflies consists of soft and flexible telescoped sections which, due to a complex arrangement of muscles and sensilla, has incredible sensitivity and manoeuvrability. It enables the fly to deposit eggs, singly or in small batches,very carefully and accurately, in a suitable place for the larvae to find food. Species whose larvae feed on aphids often lay eggs in or near aphid colonies. Those species with plant-feeding larvae usually lay eggs on particular host plants. Identifying the plant on which a female is egg-laying can, therefore, provide a useful clue to the identification of species within some genera (*e.g. Cheilosia*, *Merodon*, *Eumerus*).

Hoverfly eggs amongst aphids.

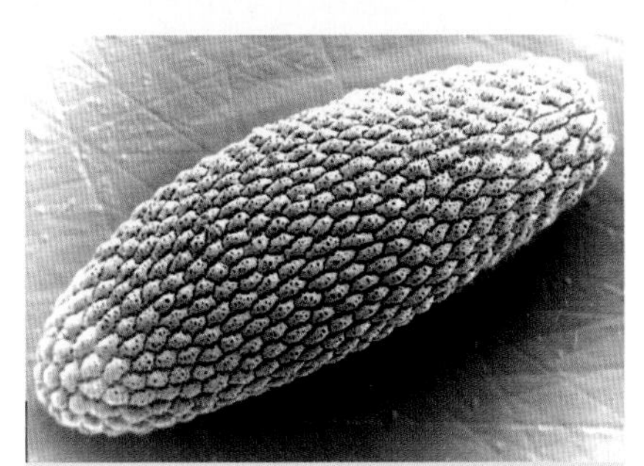

Egg of *Cheilosia vernalis* through a Scanning Electron Microscope.

Hoverfly eggs are usually somewhat elongate-oval in shape with one end wider than the other and often slightly curved. They are typically cream-coloured or pale yellow with a sculptured surface of fine pits and ridges or net-like patterns. These patterns can be distinctive and, potentially, used to identify a species, but very high magnification is required for them to be fully appreciated. When seen through a Scanning Electron Microscope the patterns are often quite beautiful. Larger species generally lay larger eggs; those of *Volucella* are around 2 mm in length, those of *Syritta* only 0·5 mm.

In the few cases where it has been observed, eggs hatch after quite a short period, typically no more than 5 days, but this is highly dependent upon temperature.

Larvae

Hoverfly larvae are typical of many dipteran larvae, being maggot-like, without legs or a head capsule. They are often somewhat wider at the rear end and narrow towards the head. Fly larvae usually have two prominent breathing tubes at the rear end: the **posterior spiracles**. In hoverflies, these are fused into a single structure: the **posterior breathing tube**. This is a characteristic of the hoverfly family which enables their larvae to be distinguished from most other fly larvae.

Larvae pass through three stages, or **instars**, during their growth. At the end of each stage they shed the outer skin so that the next stage can grow larger. The first two instars normally last just a few days. However, the third instar can last for weeks, months or even years, depending on the species. In many ways, the third instar larva **is** the hoverfly. This is the stage that does most of the feeding and its requirements largely determine the habitat and ecological importance of the species.

The ways in which hoverfly larvae live and feed are amazingly diverse and is reflected in their shapes and specialisations. The larvae of most British flies are poorly known and for some families they are virtually unknown. However, this is not true of hoverflies, since the larvae of about two thirds of the species have been described and the biology of around half is known in some detail. Keys to these species are included in Graham Rotheray's *Colour Guide to Hoverfly Larvae* (1993)*, but there are still plenty of species that require detailed study.

Larva of *Epistrophe* amongst aphids showing the characteristic posterior breathing tube.

*Out of print but a PDF can be downloaded from the Dipterists Forum website (see *page 303*).

Aphid-feeding larvae

Hoverfly larvae are well known for feeding on aphids and are consequently regarded as 'gardener's friends'. In reality only about 40% of British species are predators of this type and aphids are not the only prey. A few species feed on a variety of other soft-bodied insects, including coccids and psyllids, and a few feed on other insect larvae. For example, the larvae of *Xanthandrus comtus* feed on gregarious micro-moth caterpillars and those of *Parasyrphus nigritarsis* feed on the larvae of leaf beetles on willows and Alder along streams and rivers.

A hoverfly larva lifts an aphid from the surface of the plant as it feeds.

Predatory hoverfly larvae usually have an extensible and rather mobile front part of the body which they use to probe for prey. They are blind, so rely upon a combination of chemical senses and touch to locate prey. Having located its prey, the hoverfly larva pierces the skin with long thin mouth hooks and sucks out the contents of its victim's body. They have rather sticky saliva which is used to help hold the prey in place whilst they do this. Finally, the empty husk is discarded. While doing this, the larva often rears up and lifts the aphid off the leaf. It is thought that this behaviour helps to prevent alarm pheromones released by the aphid reaching the rest of the colony. It may take a first instar larva several hours to deal with a single aphid, but a fully-grown larva can consume an aphid in a minute or two.

One of the most unusual features of predatory larvae is that they are coloured, unlike most other hoverfly larvae, which are pale yellowish-white. Being coloured helps to provide camouflage, which reduces the risk of predation when they are moving around on the surface of a plant in search of prey. Many such larvae are green, often with longitudinal stripes of white, yellowish or brown blotches to break up their outline. Some of the most striking patterns are found in *Dasysyrphus*.

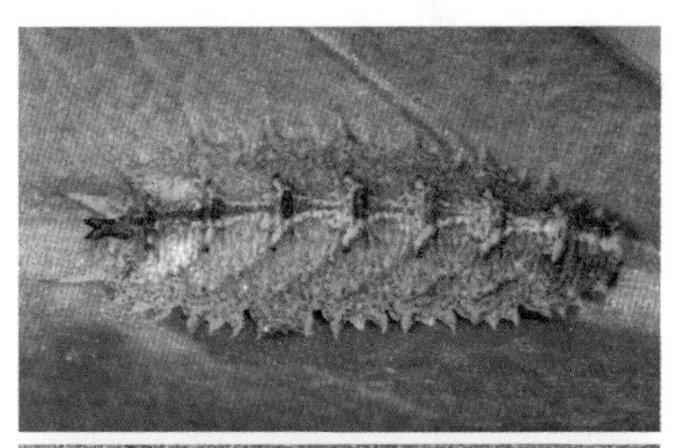

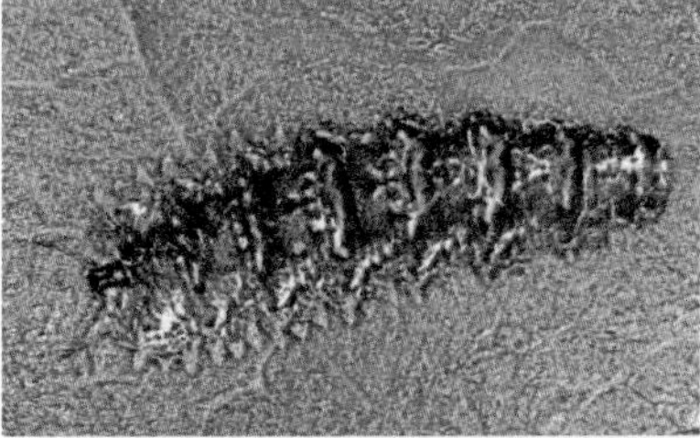

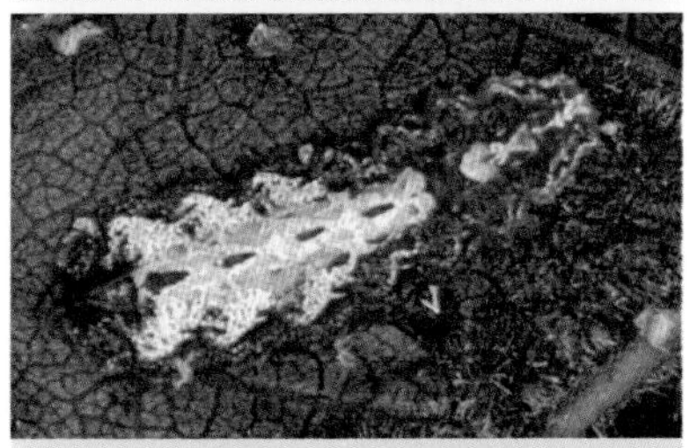

Larvae of *Dasysyrphus tricinctus* (*top*) and *D. venustus* (*middle*), camouflaged for resting on bark, and the bird-dropping mimic larva of *Meligramma trianguliferum* (*bottom*).

Galls on Lombardy Poplar leaf stems caused by the aphid *Pemphigus* and the aphids being predated by the larva of *Heringia heringi*.

These larvae feed at night on tree-dwelling aphids and rest on branches and trunks during the day. They have 'frilly edges' to break up their outline and patterns of reds, greys and browns that camouflage them perfectly against the bark. Another remarkable larva is that of *Meligramma trianguliferum*, which is a perfect bird-dropping mimic.

The larvae of the tribe Pipizini tend to specialise in feeding on aphids inside galls and leaf curls. The larvae of *Heringia heringi*, for example, are common in the galls on the stems of poplar leaves induced by the aphid *Pemphigus*. These larvae have less need of camouflage patterns because they spend much of their time hidden away inside galls – although some Pipizine larvae are green.

Larvae in the nests of ants

The larvae of a number of striking black-and-yellow hoverflies belonging to the genera *Chrysotoxum*, *Doros* and *Xanthogramma* are believed to feed on ant attended root aphids. The larvae of *Pipizella* occur in the same situation. This has been little studied: for example, it is not known how they get into an ants' nest; whether the female hoverfly enters the nest to lay eggs or whether a larva makes its own way in; and why the ants, which normally protect the aphids they farm from parasites and predators, do not kill the hoverfly larvae.

The bizarre hemispherical larva of *Microdon analis* in an ants' nest.

Doros profuges is listed as a priority species under the UK Biodiversity Action Plan and a good deal of work has been carried out to try and find out more about it. This work was prompted by a rather obscure 19th Century publication, which suggests that the larvae may be associated with ants nesting in wood, the most likely candidate being *Lasius fuliginosus*. However, the current research has had little success so far in understanding the ecology of this species.

The larvae of *Microdon* are highly specialised ant predators which feed on the eggs and brood of the host. The larvae are quite remarkable, being hemispherical and looking more like a woodlouse. Indeed, they were initially classified as molluscs in the 19th Century! The shape is probably an adaptation to protect them from the ants and they are heavily armoured to avoid bites and stings (see *page 272*).

Larvae in the nests of social wasps and bees

Volucella rank amongst our largest hoverflies and, with the exception of *V. inflata*, their larvae live in the nests of bees and wasps. *V. bombylans* larvae usually live in bumblebee nests, and the larvae of *V. inanis, V. pellucens* and *V. zonaria* live in the nests of social wasps. The larvae are mainly scavengers in the bottom of the nest cavity, feeding on dead workers and larvae, dropped food and the larvae of other insects that inhabit these nests. *V. inanis*, however, feeds directly on the wasp grubs and its larvae are flattened so that they can fit into a cell in the comb beside their victim. Larvae of *V. pellucens* have also been recorded feeding on moribund wasp larvae in the combs of abandoned wasps' nests late in the season.

Female *Volucella* have to enter the host nest to lay their eggs. Bumblebees sometimes react aggressively and it has been shown that female *V. bombylans* react to being stung by laying their eggs immediately. Observations of *V. pellucens* at nest entrances suggest that the wasps take no notice of them and they can freely enter and leave the nest.

Hoverfly larvae leave the host nest cavity in the early Autumn, when the nest is abandoned by the wasps or bumblebees. They pupate nearby: in soil if the nest is underground, or in debris in a tree hole in the case of tree-cavity nests. *V. inanis* quite often use wasp nests in buildings and consequently it is not uncommon for larvae to turn up in houses.

Larva of *Volucella inanis* in a wasp nest.

Plant-feeding larvae

Britain's largest genus of hoverflies, *Cheilosia*, together with *Eumerus*, *Merodon* and *Portevinia*, have larvae that feed in the roots, stems or leaves of plants. *Cheilosia grossa* and *C. albipila* are typical examples whose larvae mine the stems of thistles, particularly Marsh Thistle. Adults of both these species fly very early in the season (late March to April) and lay their eggs on thistle rosettes just as the plant starts to grow. The larvae of *C. grossa* may kill the growing point of the thistle when they start to feed, causing the plant to become multi-stemmed. Mature larvae mine the centre of the stem base, leaving a hollow tube. The larvae are fully grown by July or early August, at which time they exit the thistle by chewing a hole through the side of the stem and pupate in the soil surrounding the plant base.

Adults of *Cheilosia grossa* and *C. albipila* are often overlooked because they fly so early in the year. It is, therefore, much easier to record these species by searching for their larvae in thistle stems (see *facing page*). Mature larvae are relatively easy to tell apart because their posterior breathing tubes have characteristic shapes. Unlike adults, finding larvae is not dependent upon good weather, so a wet day in July can profitably be spent splitting thistle stems! This technique has shown that these species are much commoner and more widespread than records of adults suggest, especially in the uplands.

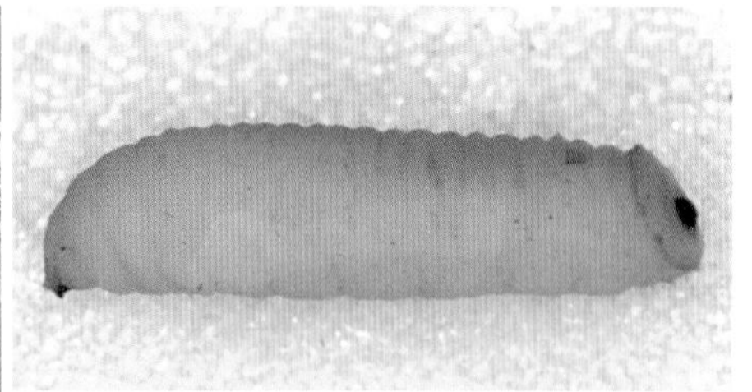

Portevinia maculata larva (*top*), Ramsons flowers (*left*) and an adult male (*above*) sitting on a leaf of the plant.

Recording *Cheilosia albipila* **and** *C. grossa* **by searching for larvae in thistles**

Opening up the rosette of a Spear Thistle.

The larvae of both species feed in the base of the stem of Marsh Thistle and those of *C. grossa* in other thistles such as Spear Thistle (but not in Creeping Thistle). They are easily found during May, June and July, or even August in the north, by splitting the base of thistle stems and looking for those that are hollowed out. Experience has shown that both species can be found in the same patch of thistles.

A good, hefty boot applied to the base of the stem will knock the plant over and open up the rosette. Insert a suitable implement and split the rosette and the base of the stem apart.

A *Cheilosia albipila* larva in situ in its tunnel after the stem has been split.

A clasp knife will do the job, but a small-bladed, sturdy gardening trowel or fork is probably better. It should be immediately obvious whether the plant has been tunnelled. An unaffected stem is filled with green, solid pith; a tunnelled stem is hollow and usually quite stained within by obvious deposits of brown frass. When you find a tunnel, split the stem upwards until you find the larva. Late in the season, you may find that the tunnel is no longer occupied – in which case it is worth a quick dig around the roots to look for mature larvae or puparia.

The larvae of these two species are quite easy to tell apart by the characteristic shapes of the posterior breathing tube (PBT). *C. albipila* is white and the PBT is parallel-sided and has a 'dagger point' projecting between the tips of the two fused hind spiracles. *C. grossa* is a dirty brown colour and the PBT is broad and blunt with a flange either side.

Other species of *Cheilosia* that occur in thistles are generally similar, but lack the diagnostic 'dagger point' or 'flanges'. Larvae of *C. fraterna* and *C. proxima* favour the more slender growth and side branches higher up on the plant, but little is known about others such as *C. mutabilis* and *C. cynocephala* (the latter of which is thought to be restricted to Musk Thistle). You may also find other insect larvae, such as caterpillars of the Frosted Orange moth *Gortyna flavago*.

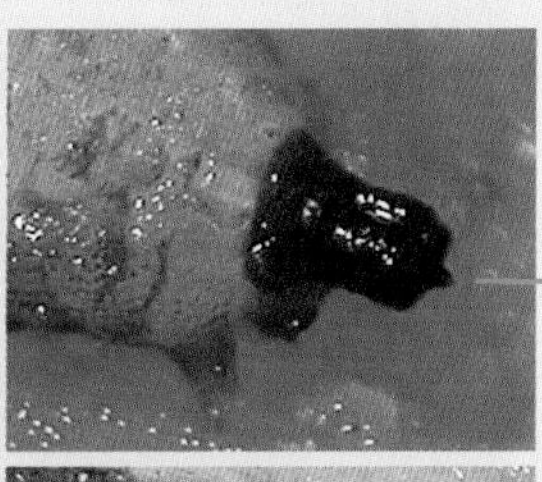

Larva of *Cheilosia albipila* and a close-up showing the characteristic 'dagger point'.

Larva of *Cheilosia grossa* and a close-up showing the characteristic flanges on the broad, blunt PBT.

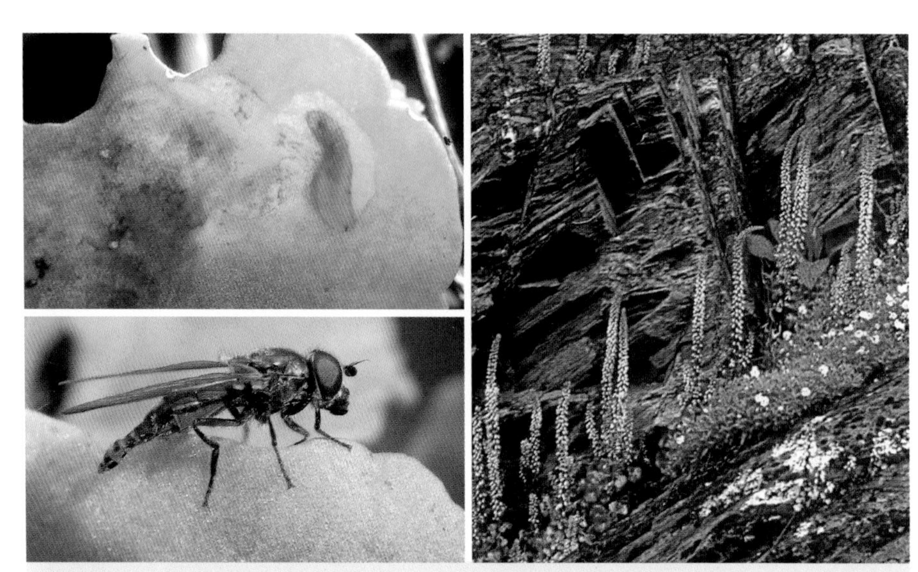

Cheilosia semifasciata larva in a leaf-mine on a Wall Pennywort leaf (*top left*), the plant growing in profusion on a slate rock face in North Wales (*right*) and an adult female of the fly (*bottom left*).

The common *Cheilosia albitarsis* and its close relative *C. ranunculi* feed on the roots of Buttercups. In these cases the larvae live on the outside of the root.

Cheilosia semifasciata is relatively unusual for a hoverfly in that its larvae are leaf-miners in Wall Pennywort in North Wales and Orpine in southern England. Since a single leaf does not provide sufficient food for a larva, they have to move from one leaf to another several times during their growth. They make quite conspicuous blotch mines, especially once the leaf starts to die, and as these blotches would be easily noticed by hungry birds, the larva chews through the petiole so that the leaf falls off as it exits the mine.

The larvae of *Merodon* and *Eumerus* mine garden bulbs, especially daffodils. It seems likely that they need the bulb to be already damaged in order to gain entry and that they do not feed directly on the plant material, but on the fungi that rot the damaged bulb.

Cheilosia longula and *C. scutellata* larvae live in the caps of *Boletus* and some other large fungi. Up to 50 larvae have been recorded in a single *Boletus* cap (they are, after all, amongst the largest mushrooms!). As the larvae consume the mushroom, it is reduced to a dark brown patch of slime and the mature larvae pupate in the soil underneath.

Saprophagous larvae

Saprophagy involves feeding on dead or decaying organic matter and is the commonest way of life for hoverfly larvae. Saprophagous larvae are filter-feeders, sucking in micro-organisms from fluids, and consequently are found in moist or wet situations. Different genera favour particular micro-habitats, as detailed in the following paragraphs.

Sap runs

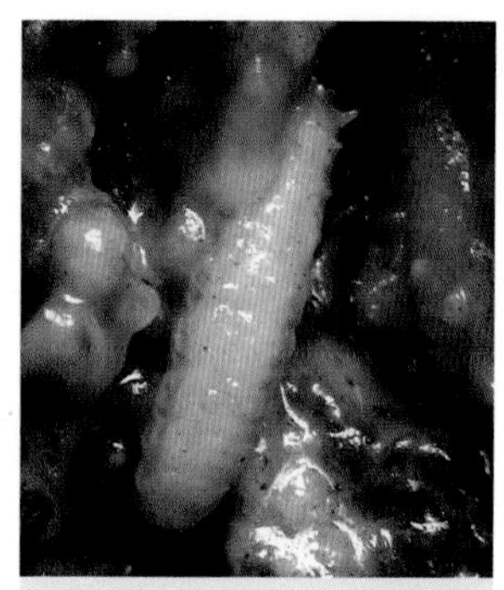

Larvae of *Ferdinandea* in a sap run.

Sap runs on trees occur when the tree has suffered some form of damage. Such runs may be transient, if the bark manages to seal itself up again, or may persist for many years, especially if micro-organisms such as bacteria become established. They are sometimes big and obvious, but most are quite small and hard to detect. Whilst all trees may develop sap runs, some seem to be particularly prone, notably Horse Chestnut, elms and Yew. Sap runs are an entirely natural phenomenon and not something to worry about in terms of the health of the tree. They should not be used as an excuse to cut down 'afflicted' trees!

Sap contains lots of nutrients and provides an ideal medium for yeasts and bacteria to thrive. Consequently, sap runs often develop a yeasty, brewing smell. Although the larvae of quite a few hoverflies can be found feeding on micro-organisms in the sap in these runs, *Brachyopa* and *Ferdinandea* are the most frequent. The larvae of these species are somewhat flattened and have projections at the side of the body which give them a spiky look. Larvae can occur in considerable numbers and it is not unusual to find a mixture of sizes, suggesting that development may take more than one year. *Brachyopa* larvae seem to have an amazing ability to resist desiccation when sap runs dry out, which they tend to do from time to time.

Rotting wood

A number of hoverfly larvae live in rotting wood. Decaying stumps and roots are favoured, particularly where the rotten wood is soft and wet and has developed a porridge-like consistency. Secondary decay by micro-organisms follows the initial decay, which is usually by fungi, and can penetrate remarkable distances into the wood. Hoverflies such as *Brachypalpoides lentus*, and the genera *Criorhina* and *Xylota* are amongst the commoner inhabitants of rotten wood, but some of the deadwood rarities, such as *Caliprobola speciosa* and *Blera fallax*, may also be found.

Bumblebee-mimicking females of *Criorhina* can often be seen flying around the bases, or inside hollows, of old stumps or crawling into crevices between large roots looking for egg-laying sites. This would be unusual behaviour for a bumblebee (unless there

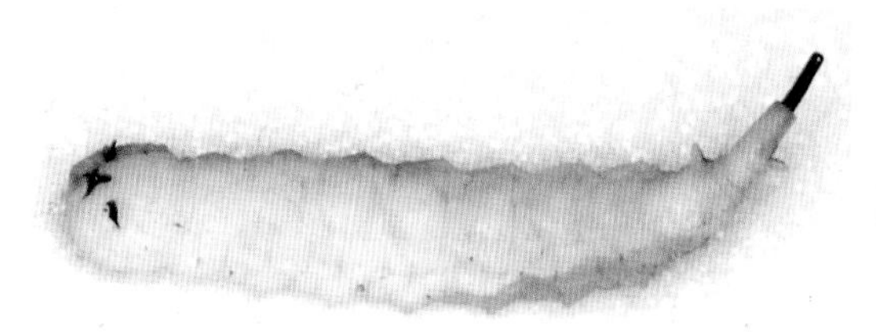

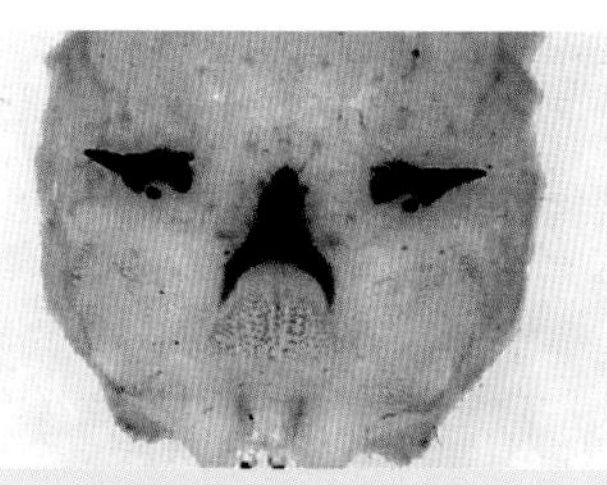

Larva of *Criorhina ranunculi* showing the large hooks developed on the thorax. These are used by the larva to get around in rotting wood and lever firmer splinters out of the way.

is an underground nest in the stump, in which case a stream of them will be going in and out of the same spot). So, if you see 'bumblebees' behaving like this they are worth checking!

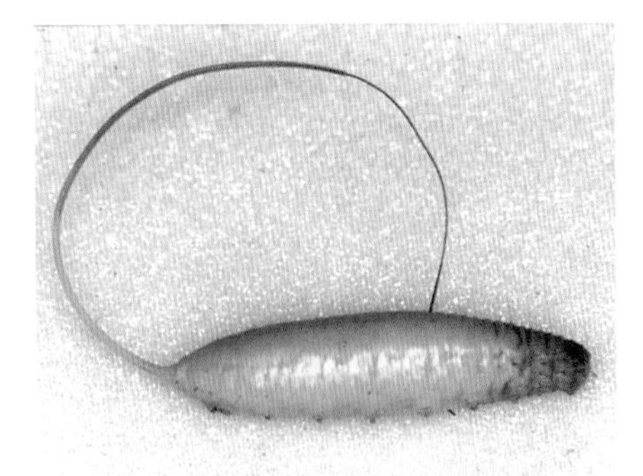

Myathropa florea larva (*top*); a rot-hole in a Lime (*bottom*).

Rot-holes

When a branch breaks off from a tree, fungi may become established in the broken stump, and, over time, a rot-hole develops. The holes in some trees, such as oaks and Ash, develop a dry rot, but in others, such as Beech, Horse-chestnut and Sycamore a wet rot develops. If the hole is positioned such that it can collect water, as well as falling twigs and leaves, a layer of wet decaying material gradually builds up. Wet rot-holes provide ideal conditions for a range of hoverfly larvae, but are especially favoured by species within the genera *Brachypalpus*, *Callicera*, *Mallota*, *Myathropa*, *Myolepta*, *Pocota* and *Xylota*.

The micro-organisms responsible for breaking down the decaying material use up the available oxygen, leaving a substrate which is often black and very smelly. This lack of oxygen makes it difficult for hoverfly larvae to breathe, and one of the adaptations seen in many of the larvae that live in such conditions is the elongation of the end of the body and the posterior breathing tube to form a telescopic snorkel which allows the larva to reach the air. Larvae in the genera *Caliprobola*, *Myolepta* and *Pocota* are described as 'short tailed', as the body is only moderately extended, whereas in *Mallota* and *Myathropa* the breathing tube is longer than the larva's body and consequently they are known as 'rat-tailed maggots'.

In some cases it may be easier to find the larvae than the adults and consequently some species are easier to record by searching suitable looking rot-holes. *Callicera rufa*, for example, was once regarded as a very rare species of Caledonian pine woods as the adults had only ever been found a handful of times. However, the Malloch Society (an organisation of Scottish Dipterists) discovered that larvae could be found by searching rot-holes in Scots Pine trees, finding them to be widespread in pine woods in northern and central Scotland and occasionally in plantations of other conifers such as Larch. This approach is potentially applicable to other seemingly rare rot-hole breeders.

Larvae living in rotting wood and rot-holes often take more than one year to develop. *Callicera* larvae (see *page 45*) typically take two years and may sometimes take as long

An Aspen log inhabited by *Hammerschmidtia ferruginea* (*left*); adult male (*top right*) and larva (*bottom right*).

as four or five. Consequently, it is not unusual to find very different sized larvae living together in the same cavity.

Under bark

When a tree falls, or a large branch breaks off, the sap under the bark decays and forms a wet, pungent layer which is full of food for hoverfly larvae. This is only a temporary condition because once the bark starts to crack and peel off it dries out. Logs therefore only provide suitable conditions for a few years. Hoverfly larvae found in this situation include the genera *Chalcosyrphus*, *Hammerschmidtia* and *Sphegina*. *Sphegina* favour logs in wet places, such as wet woodlands whilst *Chalcosyrphus eunotus* is particularly associated with partially submerged logs in wooded streams.

Other wet situations

A wide range of hoverfly larvae are associated with accumulations of wet, rotting vegetation of one sort or another, including compost heaps, dung and manure piles, and accumulations of decaying material in ponds and ditches. These larvae filter-feed on the micro-organisms responsible for the decay. Some of the larvae that live underwater are of the 'rat-tailed maggot' type (*Eristalis*, *Eristalinus*, *Anasimyia*, *Helophilus*, *Lejops*, *Parhelophilus*) and are often found in the decaying vegetation in the bottom of reedbeds

A corner of a small garden pond full of rotting leaves – ideal egg-laying conditions for *Eristalis* and *Helophilus* (*left*); *Eristalis tenax* larva (*right*).

and similar situations. *Eristalis tenax* is the archetypal 'rat-tailed maggot'; its breathing tube can be several times the length of the body when fully extended. *Sericomyia* also have long-tailed larvae that are found in more acid waters, such as in peat bogs. *Eristalinus aeneus* is found in pools near the sea-shore where decaying seaweed accumulates; although in other parts of Europe it is not a coastal species and is found in manure heaps and farmyard slurry pits.

The larvae of some genera within the tribe Chrysogastrini (*e.g. Chrysogaster* and *Melanogaster*) employ a highly-adapted strategy for breathing. Their posterior breathing tube is a stout spike which they use to penetrate the stems and roots of emergent plants to reach the air passages within, enabling them to sit and filter-feed underwater, out of harm's way.

Rotting vegetation
Some of the more terrestrial species, such as those in the genera *Syritta* and *Neoascia*, live in piles of wet decaying material including compost, farmyard manure or silage.

Rhingia campestris larvae live in cow dung. The female attaches her eggs to grass blades above cow-pats and when the larvae hatch they drop onto the surface and burrow in. The cow-pat needs to be fairly fresh because once it has developed a crust the larvae may not be able to penetrate. However, as this species is abundant in parts of the country where cows are scarce or absent (such as in woodlands in arable areas of the East Midlands and the fens), it must be able to utilise larval media other than cow-pats.

Pupae
The pupa is formed within the larval skin of the final instar so, technically, what you see is a '**puparium**' with the actual pupa hidden inside. The pupal stage may last from a week or two to several months, depending on the species. Hoverfly larvae usually pupate near the larval habitat and, if the pupa is the over-wintering stage, it is generally well hidden in soil or leaf-litter.

Migration
Migration in insects is somewhat different from the behaviour seen in migratory birds insofar as it is not the same individual that travels north in Spring to breed and returns to warmer climes for the Winter. Many insects, such as Red Admiral and Clouded Yellow butterflies, arrive and breed locally in Britain, but it is their progeny that return south (although it is an open question as to how many successfully complete the return trip).

There are a few hoverflies that migrate and do not seem to be permanently resident in Britain. One of the commonest is *Scaeva pyrastri*, which is frequent in some years but hardly seen at all in others. Most other migrants are species that are normally resident, but in some years they are augmented by immigration from continental Europe. Hoverflies that are well known for this include *Episyrphus balteatus*, *Syrphus vitripennis*, *Eupeodes corollae*, *E. luniger*, *Sphaerophoria scripta*, *Helophilus trivittatus* and various *Eristalis* species. They are sometimes encountered arriving in abundance on the south or east coasts. Occurrences of large numbers of black-and-yellow-striped hoverflies turning up on bathing beaches have led to press reports of 'plagues of wasps'.

Migratory species are capable of crossing considerable stretches of ocean. Evidence of this comes from a Malloch Society study in which traps placed on North Sea oil platforms caught hoverflies in some numbers.

Some hoverfly species have only been recorded very occasionally in Britain and are regarded as 'vagrants'. They have often occurred in locations that are also well known for their vagrant birds, such as Fair Isle, North Norfolk and along the south coast. Examples of such species are *Scaeva albomaculata*, *Helophilus affinis*, *Eupeodes lundbecki* and *E. lapponicus*.

Didea alneti, which can be extremely abundant in northern Europe, is an interesting case. It is very rare in Britain and can occur at a locality, sometimes breeding for a year or two, before dying out. The last known occurrence was in a conifer plantation in the south of Northumberland in the late 1980s. The species is believed to be irruptive, only arriving in Britain following a particularly good breeding season in Scandinavia. It has unusual abdominal markings which are often green or turquoise-blue rather than yellow.

Scaeva pyrastri – a frequent migrant to Britain.

Polymorphism and other colour variations

A number of hoverfly species show markedly different colour forms (polymorphism) within their population. This is most obvious amongst species that mimic bumblebees. *Merodon equestris* (*p. 222*), *Volucella bombylans* (*p. 244*) and *Criorhina berberina* (*p. 262*), for example, all have two or more colour forms which mimic the patterns of different groups of bumblebees. *M. equestris* is perhaps the most extreme example, with at least four (and, according to some authors, anything up to seven!) named varieties, but many individuals have intermediate characters and it is therefore difficult to name them consistently.

The sexes of some species, such as *Criorhina ranunculi* (*p. 260*), *Eristalis intricaria* (*p. 206*) and *Leucozona lucorum* (*p. 112*), have different colour patterns (*i.e.* they are sexually dimorphic) and there are others where there are morphological differences between the sexes (*e.g. Cheilosia pagana*, where the orange antennae are much bigger and more obvious in females than in males – see *page 167*).

'Brood dimorphism' is a difference in appearance between the spring and summer broods of a species. This is particularly prevalent in those species of *Cheilosia* that have two broods per year; in this instance, the spring brood is typically larger and longer haired than the summer brood.

Many yellow and black/brown hoverflies are very variable in the extent and brightness of their markings. This has been found to be related to the temperature at which they were reared. For example, in the laboratory *Episyrphus balteatus* larvae reared in cool conditions produce dark coloured adults, whereas those raised at higher temperatures are brightly coloured and with reduced black markings. This tendency is also evident in the field: adult *E. balteatus* emerging early in the year (having developed during the cooler winter months) tend to be much darker in colour (*p. 139, bottom*) whilst mid-summer individuals tend to be brightly coloured (*p. 139, top*). This is believed to aid temperature regulation because a basking hoverfly with more dark pigment will absorb sunshine better and warm up more quickly.

Many genera that breed throughout the year, such as *Eupeodes*, *Meliscaeva*, *Eristalis* and *Myathropa*, show similar variability and, unfortunately, this can make them more difficult to identify.

Variation in the markings of *Myathropa florea*: Compared to a typical individual from mid-May (*left*), this one from August (*right*), has noticeably brighter and more distinct markings.

Some colour forms of *Merodon equestris*

Form *validus* – thorax completely dark; 'tail' greyish or buff.

Form *narcissi* – both thorax and abdomen tawny.

Form *transversalis* – like *narcissi*, but with a broad black band across the abdomen.

Another, rather attractive example of *narcissi*, with an orange thorax and pale abdomen.

Form *equestris* – thorax yellow at the front; 'tail' greyish or buff.

Form *bulborum* – like *equestris*, but with a red/orange 'tail'.

Mimicry

Many hoverflies resemble bees or wasps, not only in their shape and colouring, but also in their behaviour. Some are extremely good mimics and even experienced recorders need to look twice. Others are not nearly such convincing mimics; for example many black-and-yellow Syrphini only vaguely resemble wasps, whilst several *Eristalis* species have only a passing resemblance to honey-bees – see table on *page 64*.

As previously mentioned, *Volucella bombylans* is a very convincing bumblebee mimic that breeds in bumblebee nests. Is it trying to fool the bees in order to gain entrance to the nest to lay eggs? This seems unlikely. The colour forms of adults bred out from larvae and pupae found in nests of a particular bumblebee are not the same. The probability that the colour form of the *Volucella* that emerges matches the host bee is no better than random chance. Its close relative, *V. pellucens*, which breeds in wasps' nests, looks nothing like a wasp; it has a black body with large white markings at the base of the abdomen, though observations suggest that it has no problem entering a nest.

If they are not trying to fool the bees and wasps, who or what are they trying to convince? The usual explanation is **Batesian mimicry**, a hypothesis first proposed by the English naturalist H.W. Bates. The theory is that a palatable species evolves colour patterns which imitate the warning signals of a noxious species, directed at a common predator. In other words, by looking like a stinging and distasteful bee or wasp, a hoverfly gains a degree of protection from predators, even though it is actually perfectly edible.

Experiments with birds such as starlings and pigeons suggest that they have abilities similar to humans in distinguishing between hoverflies and wasps or bees, *i.e.* they are just as likely to be confused and may have to look twice! Observations of a tame Spotted Flycatcher showed that it knew the difference. It would carefully remove the stings of wasps before eating them, but showed no tendency to try and do the same with a wasp mimic hoverfly.

However, there is little evidence that birds are significant predators of adult hoverflies. The predators that are most regularly seen feeding on them are other invertebrates, especially spiders, dragonflies, wasps, and other flies such as females of the common yellow dung-fly *Scathophaga stercoraria*. Indeed the solitary wasp *Ectemnius cavifrons* specifically targets hoverflies to stock its nest.

Two hoverfly predators: *Scathophaga stercoraria* (left) and *Ectemnius* sp. (right).

The Drone Fly *Eristalis tenax* and a honey-bee.

Volucella bombylans and a bumblebee.

Chrysotoxum cautum and a Common Wasp.

Finding hoverflies

Hoverflies are found throughout Britain, though with more species tending to occur in the south and east and less in the north and west. The fewer northern species do include some of our most charismatic hoverflies such as *Blera fallax*: a species of the Boreal forest that maintains a tenuous foothold in central Scotland. Northern species may occur farther south on higher ground or in habitats more typical of northern areas, such as conifer plantations.

Climate is one important factor that influences this distribution. Hoverflies tend to prefer the warmer and drier, continental conditions of south-east England than the wetter and cooler Atlantic conditions of northern and western areas. Some southern species have a more coastal distribution towards the north of their range where the influence of the Gulf Stream ameliorates the local climate.

Geology also plays a part; the underlying rocks of south and east Britain are younger, less acidic and provide richer soils that host a greater biodiversity than the older, more acidic rocks of northern and western Britain.

Land-use is also a significant factor. This can be seen in the distribution maps of many species where there is a dearth of records from some intensively farmed areas such as Lincolnshire and the East Anglian Fens. In these areas, hoverflies can be hard to find and are largely restricted to surviving fragments of habitat such as roadside verges, ditches and churchyards. Another example is that of conifer plantations, which tend to be located in northern and western areas and, although inhabited by a good number of species, they support considerably fewer species compared to extensively wooded areas of the south and east that have retained continuous tree cover over the centuries. Unsurprisingly these areas support our richest diversity of hoverflies.

Species lists from a good site in southern England may be in excess of 100 species (the richest sites are nearer 150 species). Further north, numbers drop quite dramatically and, in the East Midlands for example, a good site may support around 90 species. The further north you go, the more numbers of species tend to drop, so that sites in the highlands and islands of the far north and west may support only around 30 species.

The seasonal calendar

Many hoverflies are easy to find at flowers and you will quite rapidly develop a decent list of species that visit common grassland and hedgerow flowers. However, there are many other places you should look if you wish to extend your list. Fieldcraft is something which develops over time, and is as much about knowing what you are looking for, as it is employing general observational skills.

It is also worth bearing in mind, that as the season progresses, different search techniques can be employed to find some of the species that are not natural flower visitors. Once you start to develop a search pattern, you will find yourself looking out for specific features such as a sunlit tree trunk or highly reflective young leaves in a sunny spot. However, the best way to starting recording hoverflies is to follow the season and the flowering plants that attract them.

You can look for hoverflies throughout the year, even in mid-Winter when a few hardy species such as *Episyrphus balteatus*, *Eupeodes luniger*, *Meliscaeva auricollis* and *Eristalis tenax* will visit garden flowers such as crocuses, Winter Jasmine and *Viburnum* × *bodnantense*. A sunny day in January will often yield results; the first couple of records of this Millennium were of *E. tenax* on 1st January 2000!

Cheilosia bergenstammi on Dandelion.

The field season really starts in **March** once Goat Willow, Blackthorn and Cherry Plum come into flower. Sometimes these flowers can be alive with flies, but it is often hard work. The biggest problem is trying to pick out a hoverfly from a sea of white petals or to catch one from amongst interlocking branches. It is therefore often a good idea to focus on a single branch that can be viewed in isolation against a neutral background or the sky. Scanning rides for hovering flies can also pay dividends as there are several species that defend sunlit spots.

The commonest is *Eristalis pertinax* whose males often hover at head height or within ready net height. Others such as *Eristalis intricaria* and *Cheilosia grossa* generally fly much higher, so a telescopic net handle becomes invaluable (see *page 292*). Don't assume that every hoverfly you see is the same; eyes can play tricks and you stand more chance of finding new species by checking lots of individuals.

Low-growing Spring flowers such as Lesser Celandine and Wood Anemone can be useful lures, but are often rather disappointing. Even so, they will attract commoner species such as *Platycheirus albimanus*, *Cheilosia pagana* and *Melanostoma scalare*. Occasionally a more interesting species will be seen, such as *Melangyna lasiophthalma*. Young foliage, such as the fresh leaves of Cow Parsley, often provides sunny places for insects to sunbathe. This is an excellent way of finding hoverflies and will often yield a good selection of *Helophilus* and *Eupeodes* species. Don't overlook sunlit dry leaves and grass tussocks as some hoverflies will sunbathe on these too.

Eristalis intricaria on willow (sallow) blossom.

Dasysyrphus venustus on Lesser Celandine.

All of these situations can be checked as you look for a suitable 'honey pot' such as a flower-laden Goat Willow or Grey Willow (later in the Spring, Eared Willow in the uplands can be really good). These, and various members of the genus *Prunus*, are the most likely plants on which to find many of the more interesting Spring species, such as the bumblebee mimic *Criorhina ranunculi* and various *Melangyna*, *Parasyrphus* and Spring-flying *Cheilosia*.

Epistrophe eligans sunbathing on a fresh Spring leaf.

As Spring progresses the opportunities for finding hoverflies greatly increase as more and more plants that will attract hoverflies come into flower. These include Hedge Mustard, Wild Garlic, Dandelion, Cow Parsley, Ground Ivy, Bugle and White Deadnettle. Each is worth a look and may attract a different range of hoverflies. For example, Ground Ivy will attract the long-tongued *Rhingia campestris*, but is also a great place to look for *Heringia*, including the sub-genus *Neocnemodon*, which mimic tiny solitary bees as they flit amongst the flowers. Cow Parsley is a common plant that, at first sight, seems to attract relatively few hoverflies. However, this can often be an illusion because the hoverflies are widely dispersed and a good list of *Cheilosia,* Pipizini and Syrphini may be assembled from a large area of Cow Parsley.

May is an exciting time because diligent fieldcraft can often reveal species that many people miss. This is especially true of the genus *Brachyopa*, which do not generally visit flowers and require more specialised search techniques. Finding *Brachyopa* is an

A sunlit tree base and nearby vegetation provide an ideal location for *Brachyopa*.

art, but once mastered it will be found that species within this genus are by no means as scarce as past texts have suggested. Two species, *Brachyopa scutellaris* and *B. insensilis*, are actually relatively common and both can be found on sunlit tree trunks close to the ground. *B. scutellaris* is associated with a wide range of trees, including oaks, Ash, limes and poplars, whilst *B. insensilis* is usually found close to Horse-chestnut. *Brachyopa* are also inveterate leaf-baskers and will often be found on sunlit leaves of Sycamore and limes. Although Sycamore is often regarded as an unwanted invader, it is a really valuable tree for the hoverfly hunter.

Criorhina floccosa on Hawthorn.

Many other hoverflies sunbathe and there will be some days when a few species are super-abundant. For example, there is usually a two or three day period when the beautiful *Epistrophe eligans* is found in huge numbers, often with swarms of males around a particular bush.

From **late April** through to **mid-May** in southern England a wonderful array of hoverflies can be found. A warm, sunny, flowery woodland ride is a delight that is difficult to match. Further north, although the season is more compressed, finding hoverflies is still just a matter of finding the right flowers and judging whether the day is warm and calm enough to generate hoverfly activity.

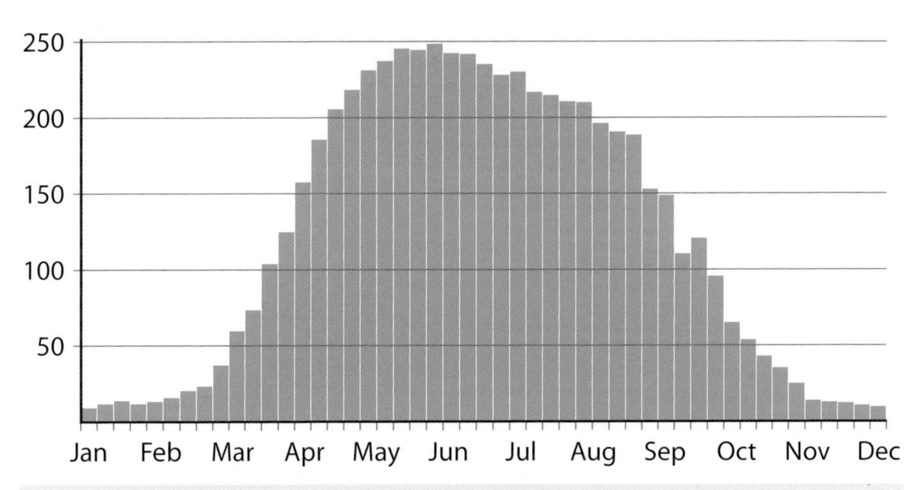

Number of species recorded each week between 1985 and 2010 by the Hoverfly Recording Scheme.

Mid to **late May** in southern England is a time of plenty when the hedgerows and woodland edges are a mass of Hawthorn flowers. These are great lures for hoverflies but really hard to work – there is so much blossom and the quick flitting hoverflies are widely dispersed. This is the time to look for species such as *Criorhina asilica* and *C. floccosa*, as well as more unusual species such as *Brachypalpus laphriformis*. Hawthorn flowers (probably a fortnight or so later) are also good lures in northern England and Scotland. However, there is possibly a better one: Rowan. The flowers of Rowan can be an immensely effective lure for a wide range of Syrphini and will also attract bigger species such as *Sericomyia silentis* and especially *S. lappona*. Rowans are also good trees to search for *Sphegina*, including the brightly coloured forms of *S. sibirica*. You will need a long net handle here too!

Once the Cow Parsley dies away and Hogweed comes into flower the Summer is upon us, and the associated hoverfly fauna is there to be found. However, although Hogweed was once a fantastic lure in southern England it is rarely productive today. The reason for this is not fully understood, but it is noteworthy that Hogweed is still a great lure in wetter parts of northern and western Britain. It may be that recent drier conditions makes the plant less nectar-rich and therefore less attractive to hoverflies. Alternatively, perhaps there are just fewer hoverflies. Recent analysis of trends in occurrence suggests that declines have been particularly pronounced in south-east England. Whatever the

Many hoverflies, mainly *Syrphus*, feeding on an Angelica umbel.

reason, it is worth keeping an eye open for other lures such as Upright Hedge-parsley and flowering shrubs such as Dogwood and Wild Privet, both of which can yield many hoverflies, especially species such as *Volucella bombylans*, *V. inflata* and *V. pellucens* together with all of the *Criorhina* species.

May and **June** are when the greatest number of hoverfly species can be found. A really warm, sunny day (but not a dreadfully hot one) will yield 30 or more species from a good site. When looking, time of day is an important factor. The number of hoverflies peaks late morning, dropping off around mid-day during hot weather, before picking up again in late afternoon. Good numbers can often be found in late evening sunshine, making the most of the sun's last few rays – so all may not be lost if you can't get out of work until late. Surveying hoverflies at this time may also yield useful information because there is comparatively little known about hoverfly activity at this time of day.

It is also worth bearing in mind that, although some hoverflies may appear to be very rare, this could be due to the fact that they have only a short emergence period restricted to just a few days each year. This seems to be the case with *Brachyopa* since when you get a productive day you will often find *B. scutellaris* at every site you visit. The same seems to hold for *Myolepta dubia* which, in Surrey, was found at four disparate locations on the same day, and yet was otherwise very scarce. It is therefore worth making an effort if you think you have hit a period of peak emergence.

The months of **July** and **August** are great times for recording hoverflies as this is when they are at their most plentiful. Resident populations are augmented by migrations of common species such as *Episyrphus balteatus*, *Syrphus vitripennis* and *Scaeva pyrastri*. There are also lots of flowers for them to visit, Angelica, ragworts, Water Mint and Fleabane being particularly favoured. For these reasons, if a rich site is worked assiduously a species list well in excess of 30 can be assembled, providing you are able to check the identity of the *Cheilosia* and *Sphaerophoria* that are often very numerous at this time of year.

Hoverfly numbers drop remarkably rapidly by the end of the third week in **August**, but they can still be plentiful well into **September** and **October**. The number of species recorded by the Recording Scheme for this time of year is not greatly lower than mid-August, but the number of individuals is considerably less. Although the flowers they like are generally dying off and there are fewer nectar sources, Ivy can be very productive in the Autumn.

Myathropa florea on Ivy flower.

Where to look for hoverflies

Hoverflies can be found almost anywhere, from urban gardens through to the seaside esplanade, and there will be occasional surprises. Choosing a site to find hoverflies also involves an element of fieldcraft. You need to know the sorts of habitats particular species favour in order to find them, so it really is important to know something about the animals you wish to find.

Sweeping is an essential part of the surveying process because it often yields species that might otherwise be overlooked. For example, a patch of Creeping Buttercup flowers will often yield more than just *Cheilosia albitarsis*. Here, *Lejogaster metallina*, *Melanogaster hirtella* and perhaps even *Pipiza* species may also be found. *Sphaerophoria* are best sought by sweeping on heathlands, and with a bit of effort several of the scarcer species can be found.

Many hoverflies occupy quite specialised niches. This makes them very useful indicators for conservation management and also means that they must be looked for in particular places. The best sites usually have large areas of high quality habitats but this is not an absolute rule and many important records come from small and otherwise insignificant sites. This is exemplified by many of the hoverflies considered to be old woodland indicators: the 'deadwood fauna'. Whilst top quality deadwood sites such as the New Forest, Windsor Great Park, Epping Forest and Burnham Beeches may yield a sizeable assemblage of species from rot-holes and decaying stumps and roots, many if not most of the species involved can be found at much smaller sites. For example, Old Sulehay Forest in Northamptonshire comprises only 35 hectares of woodland that supports very few over-mature trees and yet has yielded an exceptional list of deadwood hoverflies, including *Callicera aurata*, *Mallota cimbiciformis*, *Myolepta dubia* and *Xylota xanthocnema*.

Beech stumps in Mark Ash Wood, New Forest.

Deadwood hoverflies are not just confined to woodlands. For example, there are two records of *Callicera aurata* from a garden in Wolverhampton. Other great surprises were *Mallota cimbiciformis* in open grassland at Thrislington Plantation near Durham, and *Myolepta dubia* found in open countryside in Surrey adjacent to a spinney of no more than an acre in extent. In the case of deadwood hoverflies you must expect the unexpected – *Callicera rufa* being a prime example (see *page 45*).

Old Beech at High Standing Hill, Windsor Great Park.

Many other habitats support a range of hoverflies, although some species are more faithful to some habitats (such as heathlands, dry, hot grasslands, and wetlands) than others.

Although heathlands are normally associated with southern localities, the main component of heathland, Heather, is much more widely distributed. This means that whilst some heathland hoverflies such as *Pelecocera tricincta* are restricted to the truly southern heaths of Sussex, Surrey, Hampshire, Dorset and Devon, many others are more widespread. This is particularly true of species in the genus *Sphaerophoria*, such as *S. virgata*, *S. fatarum* and *S. philanthus*, which occur north into Scotland. Some, such as *Trichopsomyia flavitarsis*, are found in acid grasslands as well as on heathland, indicating that the association is more complex than just a relationship with Heather heaths.

Many other apparent heathland associates are actually more closely aligned with pines that often invade heathland. These habitats can therefore be great places to record species whose larvae are predacious upon conifer aphids. *Didea*, *Parasyrphus* and *Scaeva selenitica* are good examples, though they can also be readily found in conifer plantations.

Studying hoverflies on heathland is a special art and can be hard work, particularly as there are few nectaring sites apart from the Heather itself. The best sites are often those dissected by footpaths with a heathery verge containing yellow composites and Tormentil. Although such verges are often the most productive places to look, in late Summer many hoverflies will also visit Heather flowers. Some species are exceptionally difficult to find as adults. *Microdon analis* and the complex of *M. myrmicae/mutabilis* are noteworthy in this respect, but once their larval stages are known they are much easier to find. In the case of *M. myrmicae/mutabilis*, the two species can only be separated on larval and pupal characters and this requires careful examination under high magnification.

Dry, hot grasslands often exhibit similar characteristics to dry heathland; they are relatively warm and are favoured by hoverflies that prefer such conditions. Given the close parallels with heathlands, there is unsurprisingly a great deal of overlap between the faunas. Calcareous grasslands on chalk and limestone do, however, support a suite

A path on Chobham Common in Surrey. Open ground like this is important on heathland in providing warm areas. Flowers like Tormentil grow along the sides of such paths and are popular with hoverflies like *Paragus*.

of relatively specialised species such as *Microdon devius*, whose larvae are associated with the hill-forming yellow ant *Lasius flavus*. This association is peculiar because although most sites are on chalk downlands, there are sites in North Wales and East Anglia that are much damper and cooler. Quite why this species is not more widespread is somewhat of a mystery, especially as *Xanthogramma citrofasciatum*, which is also associated with the yellow ants' nests, occurs more widely on grass heath and even on grazing marshes.

Experiences from southern Scotland show that if conditions are good and there are plenty of flowers, then hoverflies can occur in good numbers. Devil's-bit Scabious has been found to be particularly popular and in northern and western regions of Britain; areas with this species are good places to search for hoverflies such as *Eriozona syrphoides*, *Didea fasciata* and *Sericomyia superbiens*. However, it is also worth bearing in mind that this also illustrates the difference in habitats between acid species-poor localities that support Heather and Devil's-bit Scabious compared with richer soils that are often much more agricultural.

Finding hoverflies is a much more demanding process where tall flowering plants are not immediately obvious. For example, moorlands can be seemingly barren until you find a damp streamside with Tormentil, which will yield all sorts of small hoverflies such as *Melanostoma*, *Platycheirus*, *Melanogaster*, *Lejogaster* and *Trichopsomyia*. The heather itself can be very attractive to hoverflies when it is in flower and, in particular, has been found to support large numbers of *Didea*. So too can Heath Bedstraw, which in Scotland is often an exceptionally good lure in woodland rides. The trick is to try to find flowers that are attracting hoverflies and then develop your technique from there. In some places you may have to sweep the vegetation and here there are definite benefits from retaining a number of specimens: they may look very similar in the net but microscopic examination can reveal several species.

Wetland hoverflies show similar variation in habitat affinities, with some favouring much more restricted habitats than others. For example, *Anasimyia interpuncta* seems to favour grazing marshes and some riverine wetlands, whereas its near relatives *A. contracta* and *A. lineata* are much more widespread, occurring in a wide range of locations where Bulrush grows, even in small roadside ditches. Some wetland species are even more specialised. *Tropidia scita*, which is readily recognised by the obvious triangular flange on its hind femora, is very closely associated with Common Reed and is most frequently found in coastal locations. Another example is *Lejops vittatus*, which is restricted to brackish grazing marsh colonised by Sea Club-rush and is usually found around the head of tidal estuaries.

Sericomyia silentis on Devil's-bit Scabious.

A ditch on a coastal marsh on which *Lejops vittatus* (*above*) occurs.

Many members of the tribe Eristalini (wetland hoverflies with 'rat-tailed maggots') are relatively catholic in their habitat preferences, but some such as the Bog Hoverfly *Eristalis cryptarum* and *E. rupium* are more specialised.

The adults of many wetland hoverflies, such as *Helophilus pendulus*, *Eristalis pertinax* and *E. tenax*, occur well away from breeding sites, although their ubiquitous occurrence may also suggest that they can use small patches of habitat that we overlook. This behaviour means that it is not always necessary to visit what are often thought to be the ideal places for hoverflies. Roadside verges often prove to be very productive, especially if there are lots of nectar sources. Indeed, in areas where there are large expanses of intensively managed agricultural land or heavily grazed moorland, roadside verges may provide the only suitable habitat.

Conifer plantations can also be very productive, especially in the uplands, where it is not uncommon to find rich assemblages of hoverflies. The reasons for this lie in the way that conifer plantations have changed the upland environment. Former sheepwalks often revert to heathland within woodland rides, and where there are wide margins, the resulting vegetation can be extremely floriferous, sheltered and protected from sheep grazing. Consequently, conifer plantations have plenty of lures to attract hoverflies and the hoverfly recorder.

Finally, it is worth remembering that female hoverflies need to eat pollen to provide protein for egg production and will often visit pollen-rich sources such as grasses and plantains. A list of hoverflies for any site can be greatly enhanced by sweeping a verge with flowering Ribwort Plantain. Apart from the common *Melanostoma* species, numbers of *Platycheirus* will occur, especially *P. clypeatus*, *P. angustatus* and *P. manicatus*. Sweeping is also a very effective way of finding *Sphegina* in northern and western Britain, especially where Pignut abounds: occasionally three, or even all four, species in this genus can be found at the same locality, whereas visual searches may be far less successful.

A male *Platycheirus* on a grass head.

The search for *Callicera rufa*

Adults of *Callicera* are very elusive. It is thought that they spend much of their time high up in the canopy and that the only times they are to be found at ground level are when they emerge, if they need to come down to drink or when females lay eggs. Consequently, the spectacular *C. rufa*, a resident of the Caledonian pine forests of central Scotland, was recorded very few times before 1988 and was believed to be a great rarity.

In the late 1980s work by the Malloch Society showed that *C. rufa* was much easier to find by looking for larvae in water-filled rot-holes in Scots Pine. During the course of a few field seasons they were able to record it from almost all sites in the Scottish Highlands where Caledonian pine forest still exists. It was also found in suitable rot-holes in mature Scots Pine plantations and even in Spruce and Larch on occasion. More recently, larvae have been found in water filled cavities in the stumps remaining after mature plantations have been felled.

Artificial cavities can be created in stumps using a chainsaw and naturally fill up with rainwater. These are occupied very quickly, often in the same year they are made, and have proved to be a good way of finding out whether the species is present in an area.

Records of adults, however, remain very sparse and infrequent (apart from those reared from mature larvae brought into captivity).

Callicera rufa.

Artificial rot-hole created in the stump of a Scots Pine using a chainsaw.

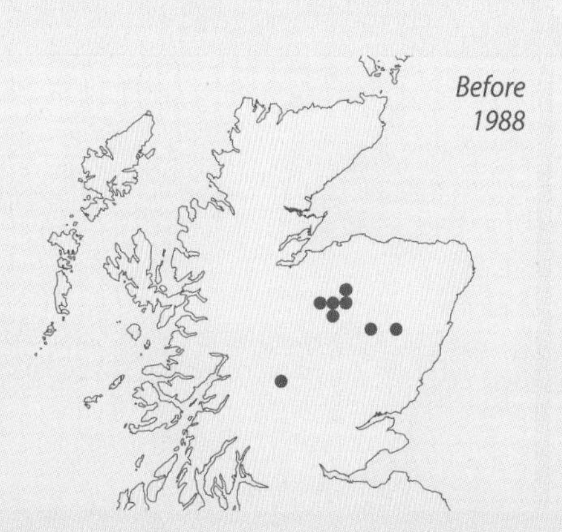

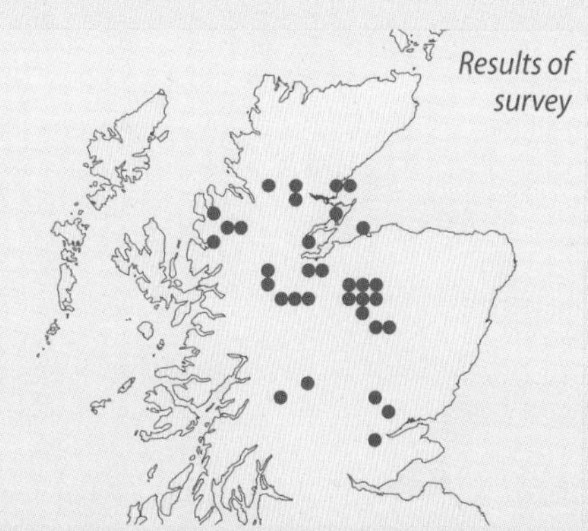

Glossary

Diagrams showing the locations of terms in **bold brown text** can be found on *pages 50–54*.

Adpressed	Pressed closely against, or lying flat.
Antenna (plural: **antennae**)	The 'feelers' on the front of the head of an insect. These bear chemo-sensory organs and allow the detection of chemical stimuli such as the odours emitted by sap or fungi – see *page 51*.
Aphidophagous	Feeding on aphids.
Apical	In the direction of the apex. Hence, that part of an appendage lying nearest its tip and furthest from the point of attachment – also **distal** (opposite of **basal**).
Aquatic	Living in water. Strictly, this covers any type of water, fresh or salt, but when applied to Diptera it usually means freshwater.
Arista (plural: **aristae**)	Appears as a bristle arising from the surface of the third antennal segment. It is actually the remnant of antennal segments and can show signs of segmentation. It can be bare, clothed in short hairs ('**pubescent**') or long hairs ('**plumose**') – see *page 51*.
Basal	In the direction of the base. That part of an appendage nearest to the point of attachment (opposite of **apical**).
Cell	An area of the wing membrane bounded by the wing margin and/or veins. Wing cells are named after the vein on their anterior side and given lower-case abbreviations (*e.g.* 'cup').
Chitin	The tough, protective, semitransparent substance that forms a hoverfly's body, wing veins *etc.*
Compound eye	Eyes made up of large numbers of ommatidia. In the Diptera, the compound eyes normally occupy a large part of the head.
Costa	The main vein forming the leading edge of the wing.
Coxa (plural: **coxae**)	The basal segment of a leg; the part attached to the thorax.
Cross-vein	Short veins connecting the length-wise veins and their branches.
Cuticle	The hard, protective layer that forms the outer surface ("skin") of an invertebrate.
Dimorphic	Occurring in two distinct forms.
Dimorphism	A difference in size, form or colour between individuals of the same species, characterizing two distinct types.
Discal cell	A closed cell in the centre of the wing bordered by **M** veins and closed by cross-vein **R-M** – see *page 52*.
Distal	Farthest from the mid-line of the body. Another term for **apical**.
Dorsal	On the upper surface.
Dorsum	The dorsal surface, usually of the **thorax** – hence **thoracic dorsum**.
Dusting	A characteristic of the surface of the chitinous plates making up the body of a hoverfly. 'Dust' is actually formed by minute, flattened hairs rather like the scales of Lepidoptera (butterflies and moths) – see *page 54*.
Entomophagous	Growing in or on an insect, for example certain fungi.
Face	The plate that forms the front of the head, delineated by the antennal sockets above, the mouth opening below and, laterally, by the compound eyes.
Femur	The principal leg segment, analogous to the 'thigh', located between the **trochanter** and the **tibia**.
Frass	The droppings of plant-eating (phytophagous) insects.

Frons	The plate forming the top of the head, bordered by the compound eyes, the **ocellar triangle** and the antennal bases.
Genitalia	The copulatory organs. The shape and arrangement of the genitalia are often used to distinguish between closely related or very similar species.
Haltere	Remains of the hind wing of Diptera which has become an organ of balance.
Humerus (plural: **humeri**)	The anterior corners or 'shoulders' of the **thoracic dorsum** – see *page 50*.
Imago	Adult
Instar	The stage in an insect's life history between any two moults. A newly-hatched insect that has not yet moulted is said to be a first-instar larva. The adult (**imago**) is the final instar.
Integument	The 'skin' or outer membrane.
Jizz	The often indefinable characteristic impression given by an animal or plant, usually defined by shape or movement.
Larva (plural: **larvae**)	The immature form of an insect which is markedly different from other life-stages such as the **pupa** or adult (**imago**).
Malaise trap	A large, tent-like structure used for trapping flying insects, particularly Diptera and Hymenoptera.
Metatarsus	The most basal of the five **tarsal segments**; also called the 'basitarsus' – see *page 53*.
Microtrichia	The microscopic hairs on the surface of the wing membrane.
Occiput	The back of the head, behind the compound eyes.
Ocellar triangle	The plate on which the ocelli are located. Usually a sharply-delineated, triangular area of the **frons** – see *pages 50, 51, 54*.
Ocellus (plural: **ocelli**)	Simple eyes. Nearly always three, arranged in a triangle at the **vertex** of the head and located on a sharply delineated triangular plate – the **ocellar triangle**.
Ommatidium (plural: **ommatidia**)	One of the individual structural elements of the **compound eye** of an insect.
Oviposition	The act of laying eggs.
Ovipositor	The egg-laying structure of the female.
Petiole	A slender stalk between two structures. In this context: the stalk formed where wing veins join; which then runs to the wing margin (see, for example, the illustration at top of *page 201*).
Phytophagous	Feeding on plants.
Pleura (singular: **pleuron**)	The plates making up the sides of the **thorax**.
Plumose	With long hairs or bristles. Usually applied to the **arista**.
Polymorphic	Occurring in several distinct forms.
Porrect	Extending horizontally, not drooping.
Posterior	At the back end, towards the tail. The rearward-facing surface of a structure.
Proboscis	In Diptera this term refers to the extensile mouthparts.
Proximal	Another term for **basal**.
Pubescent	With short hairs. Often applied to the **arista**.
Pupa	The stage in a hoverfly's life-cycle, often quiescent (inactive), that precedes the emergence of the adult (**imago**).
Puparium	The pupal **integument** or shell.

Rostrum	A snout-like projection of the head.
Saprophagous	Feeding on dead and decaying organic matter.
Scutellum	The shield-shaped, posterior part of the **thoracic dorsum**.
Scutellar hairs and bristles	The marginal hairs and bristles near the apex of the **scutellum**.
Spiracle	An air inlet for the insect's breathing system. In the Diptera, the **thorax** has two spiracles on each side.
Spurious vein	See **vena spuria**.
Squama (plural: **squamae**)	Two (upper and lower) lobes at the base of the hind margin of the wing adjacent to the **halteres**.
Sternite	The plate forming the bottom, or ventral surface, of a segment of an insect's body. Usually refers to the under-side of an abdominal segment when used in keys to Diptera.
Stigma	Coloured area of the wing next to the **costa** and near the end of veins **Sc** or $\mathbf{R_1}$ – see *page 52*.
Subcosta (Sc)	The second, usually unbranched, longitudinal wing vein, posterior to the **costa**.
Sweeping	Sweeping a net gently back and forth through low vegetation.
Synanthropic	Mainly occurring in habitats created by man.
Tarsus	The apical (outermost) part of a leg, consisting of five segments (tarsal segments or 'tarsomeres'). The first segment is termed the 'metatarsus' or 'basitarsus' and the fifth carries the claws – see *page 53*.
Taxon (plural: **taxa**)	A general term for a unit of biological classification. Often used as an equivalent to species, but this is not really correct since it can refer to any level in the taxonomic hierarchy, *e.g.* a genus or a sub-species.
Tergite	The plate forming the top, or **dorsal** surface, of a segment of an insect's body. Usually refers to the top of an abdominal segment when used in keys to Diptera.
Terminal	At the end.
Thoracic dorsum	The dorsal surface of the thorax.
Thorax	The central division of the body of an adult insect, consisting of three fused segments each of which bears a pair of legs and the hind two of which bear a pair of wings, when present.
Tibia	The fourth segment of a leg, analogous to the 'shin', between the **femur** and the **tarsus**.
Trochanter	The small, second segment of a leg between the **coxa** and the **femur**.
Vagrant	An individual that wanders outside the normal range of its species.
Vein	Chitinous, rod-like or hollow tube-like structure supporting and stiffening the wings in insects, especially those extending longitudinally from the base of the wing to the outer margin.
Vena spuria	A longitudinal fold in the wing of the family Syrphidae which is chitinised along its crease. It runs between the **R** and **M** veins and crosses **R-M**. Although it looks like a vein it is not connected to the rest of the **venation**.
Venation	The arrangement of the wing veins – see *page 52*.
Ventral	On the lower surface.
Vertex	The highest point (especially of the head); the apex.
Zygoma	A sharply defined region of the face running along the eye margins; a feature of the tribe Cheilosiini – see *pages 61, 160*.

Identifying hoverflies

A few hoverflies can be readily identified by sight in the field, even from some distance. For example, the large black-and-white *Volucella pellucens*, hovering at head height in a sunbeam along a woodland path, is quite unmistakable and can be easily identified without needing to get close.

Many more species can be identified in the field with a little more practice and experience; and that is what this book is primarily about. In many cases, you are going to need to get closer, perhaps within half a metre (a foot or two) of the flower on which the hoverfly is feeding, and have a really good look to see the necessary features. For example, amongst the common, bee-like *Eristalis* you need to find out whether it has a black stripe down the middle of the face, what colour its front feet are, and so on. Providing you can get a good, close look as it moves about on the flower you may be able to see the necessary features. Close focussing binoculars can help and many modern roof-prism binoculars are capable of focussing down to 1 m or less.

Where even more detailed examination is necessary, it is best is to catch the hoverfly in a net, pick it out carefully between your thumb and forefinger and examine it closely, perhaps with the aid of a 10× hand lens (see *page 291*). Using the sorts of tips and tricks covered by this book, with sufficient experience you should be able to identify at least 100 or so of the British species this way. Once you have checked the fly over, it can be released unharmed.

At least half of the British species cannot be identified like this. You need to take a specimen home and examine it (dead and pinned) under good lighting and higher magnification (preferably, under a binocular microscope at around 10–30× magnification). The identification of many of these species goes beyond the scope of this book and to do so you will need a more detailed guide that includes keys. *British Hoverflies* (second edition) by Stubbs and Falk (2002) is the best available UK text for this purpose.

Male *Cheilosia illustrata* – One of the more straightforward to identify hoverflies.

Naming the parts

Knowledge of the terminology used when describing the anatomy of a hoverfly is important in understanding the descriptions given in the species accounts in this book.

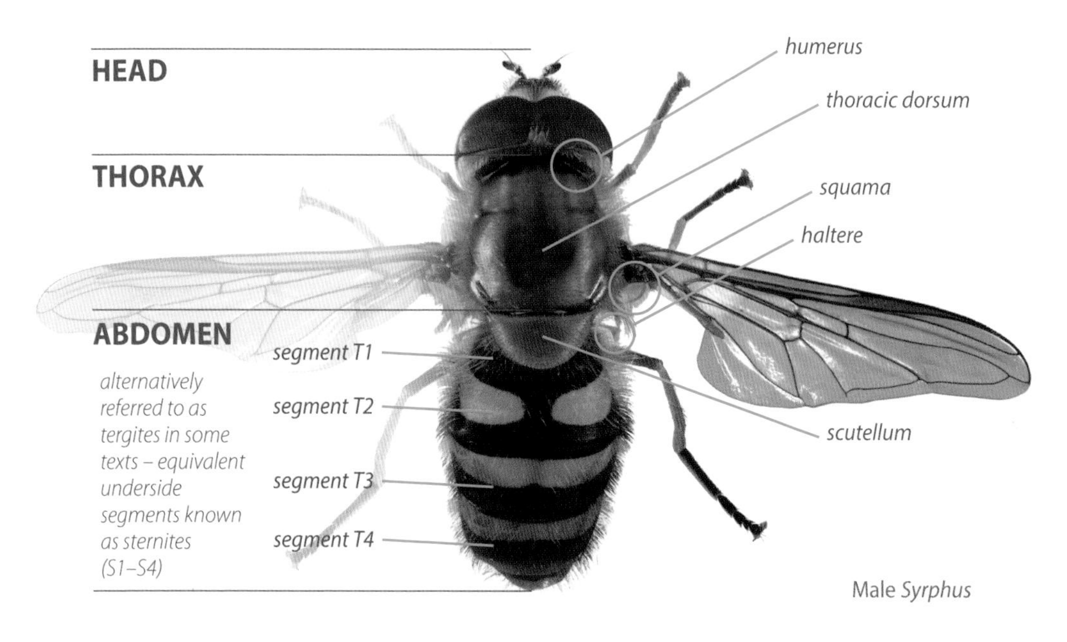

Male *Syrphus*

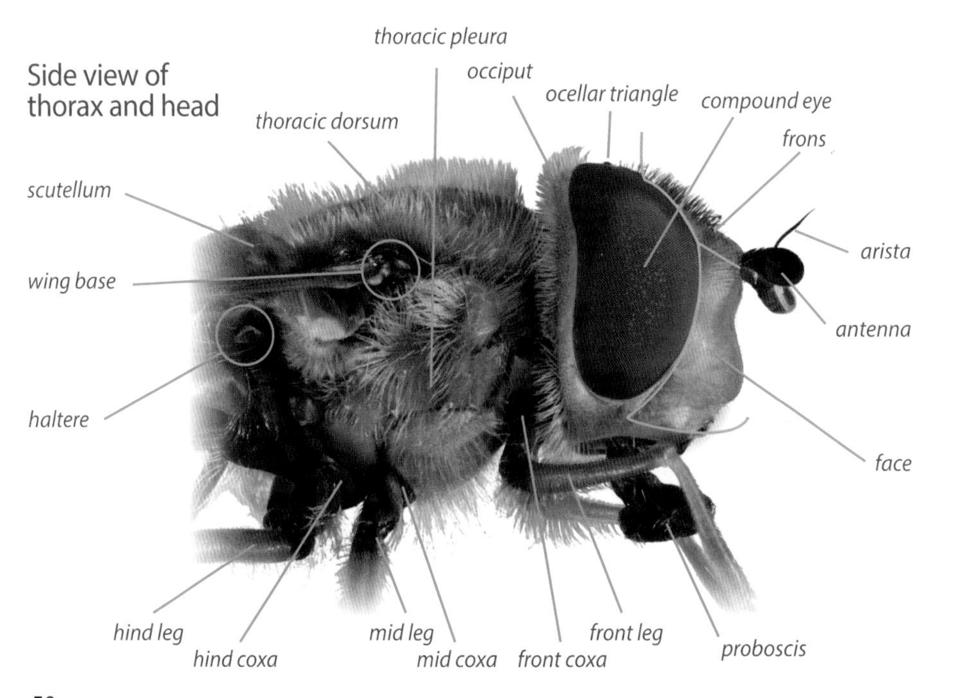

HEAD

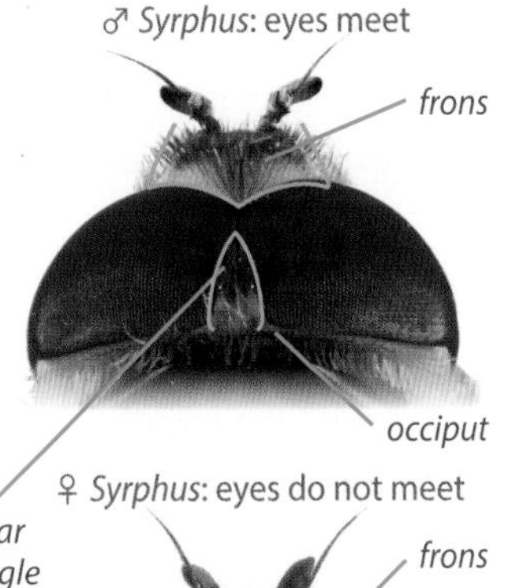

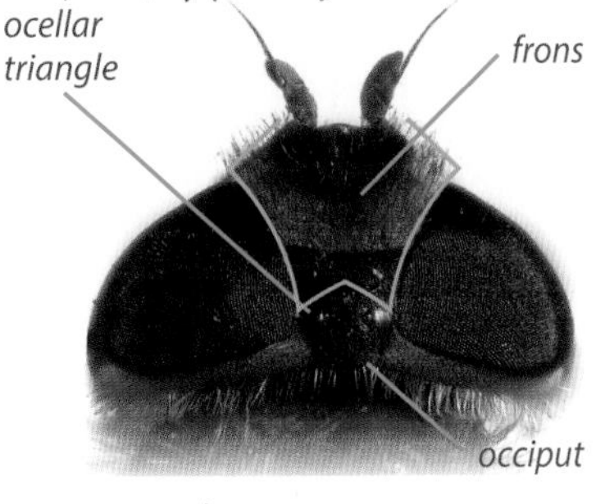

The head is dominated by large compound eyes. These generally meet at the top of the head in males, but are separated in females. Although this usually provides an easy way to sex a hoverfly, there are a few genera for which this rule does not apply (*e.g. Anasimyia, Helophilus, Lejogaster, Lejops, Microdon, Neoascia, Parhelophilus, Pelecocera* and *Sphegina*). If the eyes do meet, then you can be sure it is a male, but if they are separated the sex of the individual depends on the genus.The area on top of the head between the eyes is the **frons** (which is rather small in the males of those species where the eyes meet). Right at the top of the head there are three simple eyes, or **ocelli**, in a triangular formation, usually on a slightly raised area known as the **ocellar triangle**. The ocelli are not capable of image resolution, but they are sensitive to light and are used to measure day length and regulate a hoverfly's internal clock. A pair of **antennae** are situated below the front end of the frons. Each consists of three segments (conventionally numbered from the base outwards), with the third segment usually being the largest. It varies considerably between species, and the shape and size of the antennae is often used in descriptions. The third segment bears the **arista**, which usually arises from the top surface somewhere between the base and the middle, when it is described as dorsal. However, the arista sometimes arises from the tip, in which case it is described as apical. The aristae may be bare or hairy: if it has very long hairs so that it looks like a feather, or a TV aerial then it is described as **plumose;** if the hairs are short then it is **pubescent**; if absent then it is referred to as **bare**.

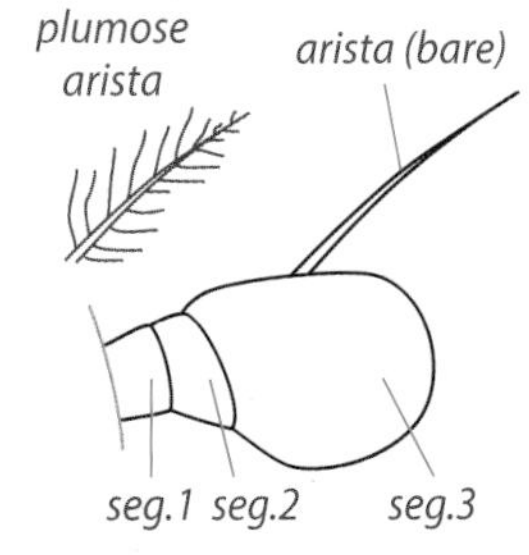

'normal' antenna segments

The **face** is below the antennae and between the eyes. It occupies the area between the base of the antennae and the mouth margin. Its colour is often useful for identifcation. The face often has a nose-like bulge or 'knob' in the middle and the presence or absence and shape of this is frequently mentioned in descriptions; to appreciate this feature you need to view the hoverfly's head in profile. The frons or face may be dusted – see *page 54*.

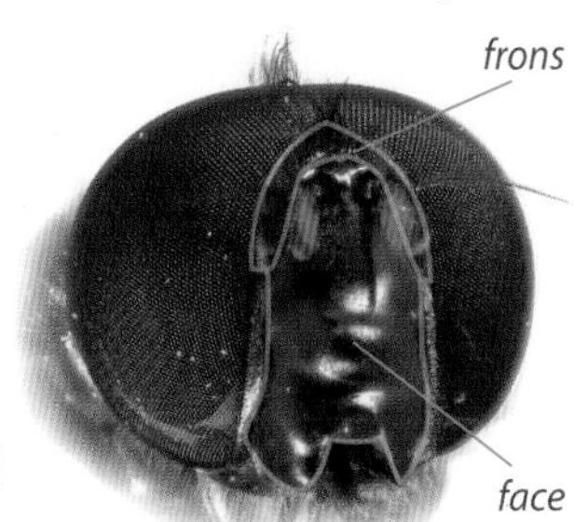

Head of *Cheilosia*

THORAX

The thorax bears the wings and three pairs of legs and also, just behind and below the wing bases, the **halteres**. The top of the thorax is the **thoracic dorsum**, the colour and pattern of which is sometimes helpful in identification. At its front corners there is a pair of swellings at the shoulders called the **humeri**. Behind the dorsum is a semi-circular swelling termed the **scutellum**. The sides of the thorax are termed the **thoracic pleura**.

Wings

The wings consist of a transparent membrane supported by a series of struts: the **veins**. The main veins run from the base of the wing towards the tip and are occasionally linked by **cross-veins**.

The naming of wing veins in flies is a complicated subject and many systems have been devised. Unfortunately, these various systems have sometimes used the same name for different veins, which can be very confusing when comparing descriptions in different books! The system adopted here (the Comstock-Needham system) is also used in *British Hoverflies* by Stubbs and Falk (2002), and most of the more recent European works. It recognises the following main veins running from the wing base (from front to back): the **costa** (C), **sub-costa** (Sc), **radial vein** (R), **medial vein** (M) and **anal vein** (A). It is thought that, in the most primitive flies, these veins had a series of branches, but that some of these branches have fused together again during the course of later evolution. As a result, seemingly strange labels such as R_{2+3}, R_{4+5} and M_{1+2} are used to refer to veins that are believed to derive from these fusions. Where a section of the wing membrane is surrounded by veins (or by veins on three sides and the margin of the wing on the other) it is called a **cell.** Cells are also given names.

The most important feature to recognise is the **R-M** cross-vein in the middle of the wing. This is always present in hoverflies and is almost always near to or at the middle of the wing. **R-M** arises from vein $\mathbf{R_{4+5}}$ and forms the outer border of the **1st basal cell**. The **R-M cross-vein**, together with the **2nd basal cell** (the one immediately below the 1st basal cell and with 3 veins arising from its outer end) and the cell below it (the **discal cell**) are the main features you need to be able to find (see *page 56*) in order to follow the descriptions.

There is often a coloured area of wing membrane between the tip of the sub-costa and the tip of vein R_1, termed the **stigma**. Some species also have a strong darkening across the middle of the wing membrane, referred to as a **wing cloud**. This shading often extends from just behind the stigma, around the R-M cross-vein and over the end of the discal cell.

The wing membrane is usually covered in tiny hairs called **microtrichia**. To see these, high magnification (around 30–40×) is needed, with the light coming through the wing membrane from behind. In many cases, microtrichia do not cover the entire wing surface, and the patterns they make are a useful character in the identification of some difficult species. The degree to which the 2nd basal cell is covered is most commonly used.

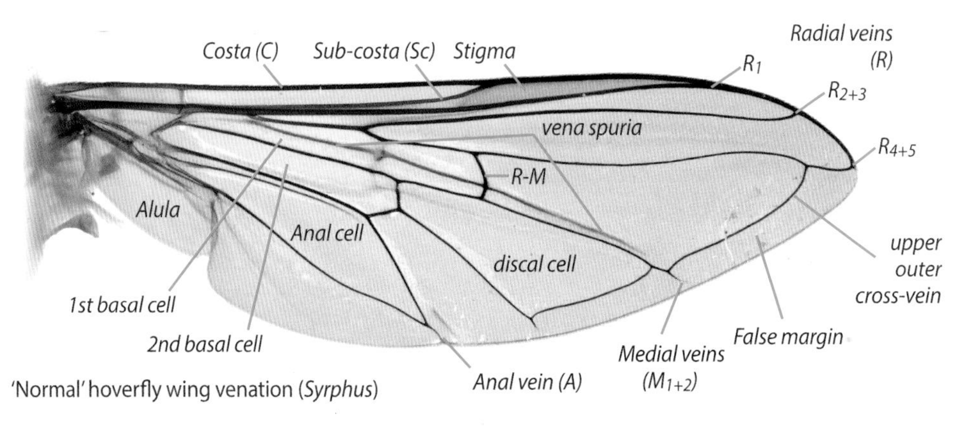

'Normal' hoverfly wing venation (*Syrphus*)

Legs

Each leg is attached to the thorax by a **coxa** and a small segment termed the **trochanter**.

The two main parts of the leg, the **femur** and the **tibia**, have the same names as the two main leg bones of a human and, like our legs, have a 'knee' joint between them. Finally, the hoverfly's equivalent of our foot is composed of five **tarsal segments** or **tarsi**. The first of these is usually the longest and is called the **metatarsus**. The last tarsal segment (seg. 5) bears a pair of claws.

When species descriptions refer to the **base** of a leg joint, they mean the part nearest to the body, so the image (left) could be described: *hind femur mainly yellow, black only at extreme base"*. The opposite is the **apex**, or '**apical**', which is the part farthest from the body.

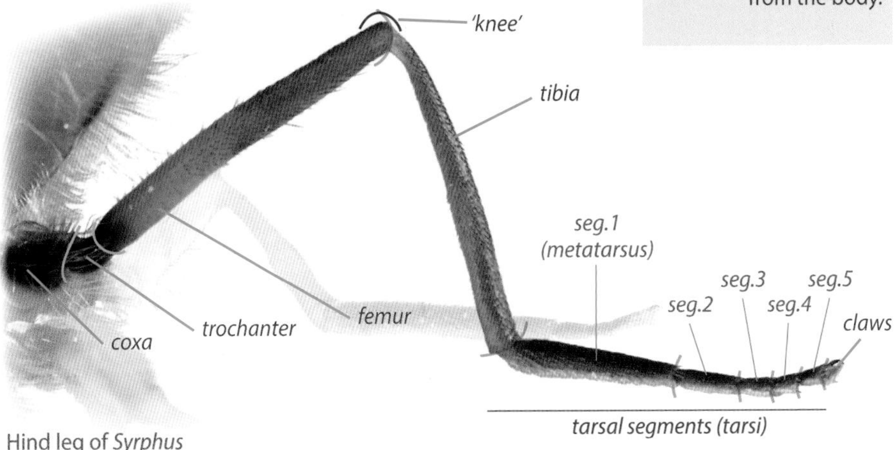

Hind leg of *Syrphus*

ABDOMEN

The abdomen is composed of a series of similar segments (or **tergites**) which are numbered from the base (where the abdomen joins to the thorax) towards the tip. The 1st segment (T1) is not usually very conspicuous and is often largely hidden by the **scutellum**. Consequently, the 2nd segment (T2) is usually the first that is obvious when viewed from above. Segments T2, T3 and T4 bear coloured markings in some species. The shape, colour and positions of abdominal markings are used a great deal for identification. The **genitalia** are positioned at the tip of the abdomen. In the male, they are often fairly obvious as a somewhat globular capsule, folded under the end of the abdomen and forming a distinct bulge in side view. The abdomen of the female usually tapers to a blunt, conical point, with no trace of a bulge, providing another way of determining the sex of a hoverfly in species where the males' eyes do not meet. Very occasionally the colour or dusting of the plates on the underside of the abdomen is used in species descriptions. These plates are termed **sternites** (S1–S4).

Abdomen of male *Syrphus*

Frons and face of a female *Melanostoma scalare* showing dusting

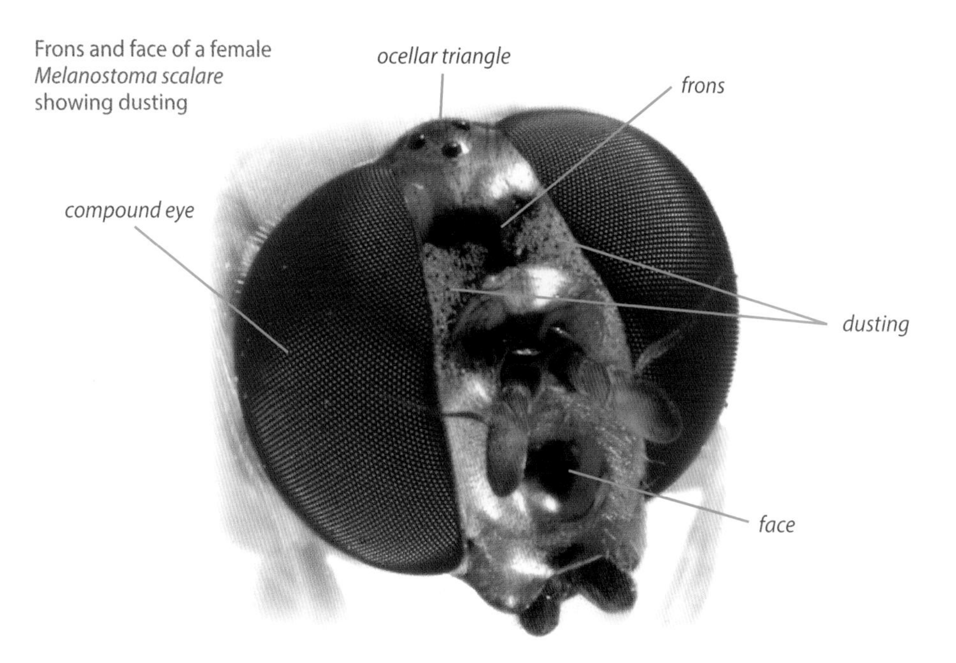

Dusting

Patterns and markings in hoverflies are of two kinds: coloured areas of the **cuticle** and patterns formed on the surface by bands of differently coloured hairs and **dusting**.

Dusting is actually formed by tiny flattened hairs. The name is quite descriptive: it looks like patches of dust on the shiny surface of the hoverfly's cuticle. The characteristic feature of dusting is that its appearance depends on the lighting. As you move a hoverfly around so that the light comes from different directions, the appearance of dusting changes. At some angles it may almost disappear, at others it may be obvious, contrasting with the shiny cuticle.

The appearance of patterns formed by bands of coloured hairs changes in a similar way according to the direction and intensity of the light. It can often be difficult to assess the colour of hairs, and fine black hairs can appear to be pale when brightly lit. This is because you are actually seeing the bright reflections off the surface of individual hairs rather than their true colour.

By contrast, markings due to coloured patches of the insect's cuticle don't tend to change in appearance as the direction of lighting changes.

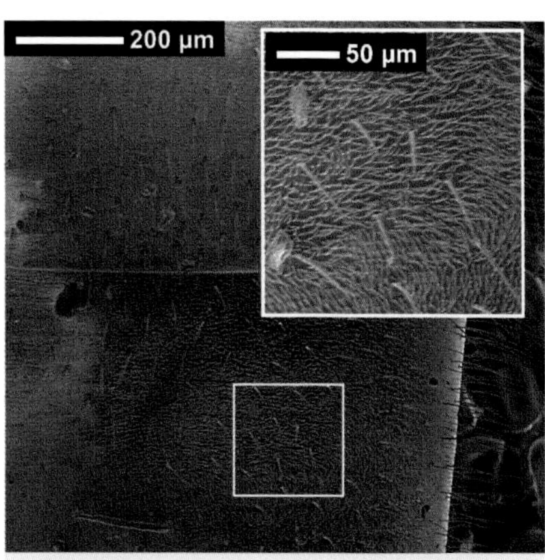

Scanning Electron Micrographs of dust patches on the abdomen of a female *Platycheirus albimanus* showing that they are composed of tiny hairs.

Guide to the tribes

The following section gives a reasonable but not foolproof introduction to the characters that are best used as pointers to identify tribes and sometimes genera or species within that tribe. It focuses on features which can be seen in the field or by using a 10× hand lens. Naturally, experience will help in speeding up this process but faced with an unfamiliar hoverfly, here is a suggested process to assist identification.

A summary of British hoverfly tribe features can be found on *page 63*.

1) Confirm that is is a hoverfly by the presence of the vena spuria (except for *Psilota* – see **2a**)

2) Establish whether the front of the thorax behind the head is visible or obscured and whether the humeri are hairy or bare:

3) **If they are obscured and/or bare:** look at additional features of the Syrphini, Bacchini and Paragini (*p. 62*)

If they are visible and hairy: look initially at the wing venation and, from there, additional characteristics of the face, antennae and aristae.

4) Once the tribe has been established go to the relevant *Guide to tribe* page to identify the genus.

1

a) Front of the thorax generally visible, making it possible to see the humeri – which are hairy

Very variable in form, and includes the majority of the big bee and wasp mimics.

Most hoverfly tribes.

b) Head concave, making it hard to see the front of the thorax, obscuring the humeri – which are bare

Includes the majority of black and yellow hoverflies mostly (except **Bacchini**) with at least some yellow on their faces.

Syrphini *p. 94* **Bacchini** *p. 72* **Paragini** *p. 92*

2

from **1a**

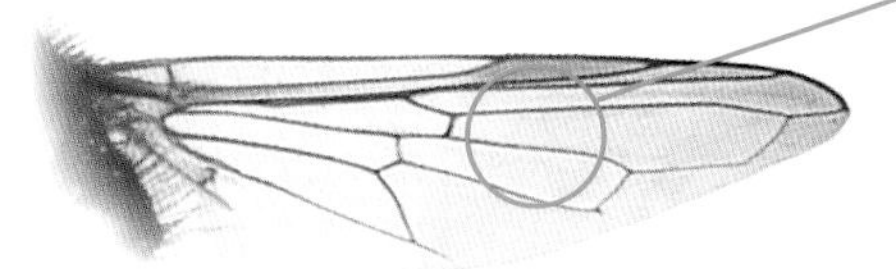

a) Vena spuria absent

The presence of a vena spuria is a feature that is found in all British hoverflies, and can be fundamental to the identification of those species that mimic or look very similar to other Hymenoptera or Diptera.

There is one exception: the absence of a vena spuria is a feature unique to *Psilota anthracina* amongst British hoverflies. Beware – great care should be taken with its identification as this rare species is easily confused with some blue-black muscid flies.

Merodontini: *Psilota* *p. 224*

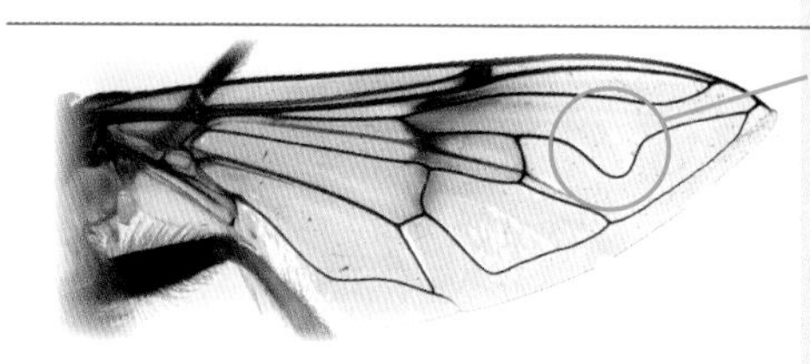

b) Strong loop in wing vein R_{4+5}

A group of mainly medium-sized hoverflies including convincing honey-bee and bumblebee mimics.

Beware – some Syrphini *e.g. Didea*, the subgenus *Lapposyrphus* (within *Eupeodes*) and *Megasyrphus* have a dip in R_{4+5}. See *page 95.* However, checking the head + humeri (see **1b**) should avoid any confusion.

Eristalini ***p. 198***
Merodontini: *Merodon* *p. 222*

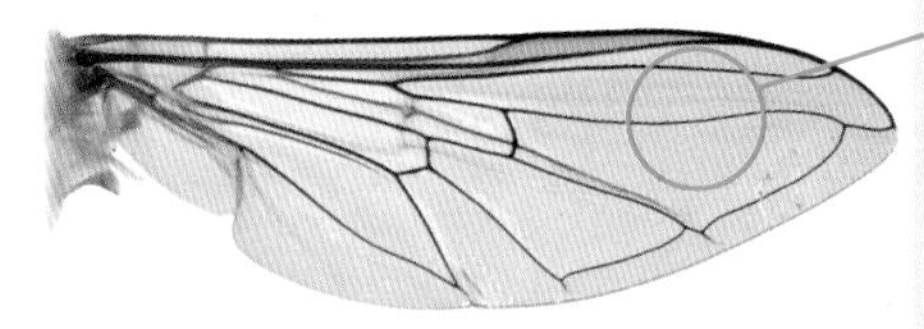

c) 'Normal' wing – no wing loop present

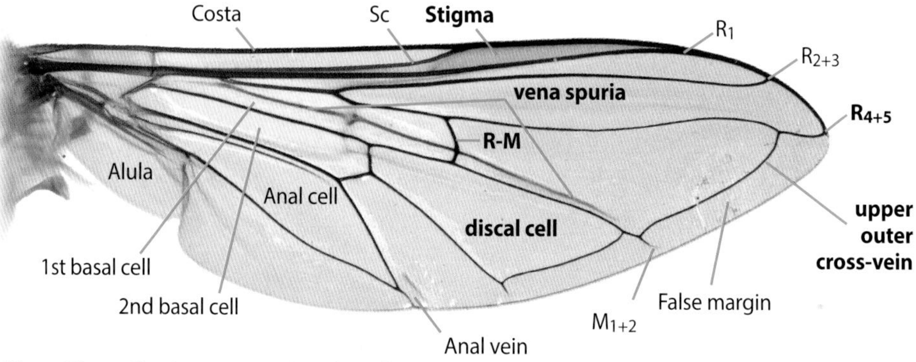

'Normal' hoverfly wing venation (*Syrphus*) (from *page 52*); features mentioned in the key in **bold text**.

3
from **2c**

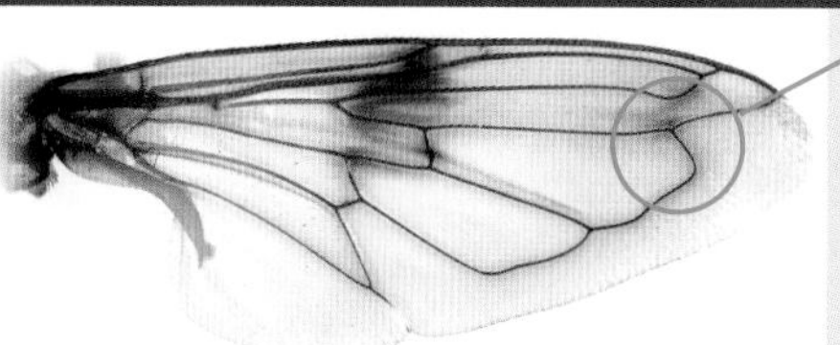

a) Upper outer cross-vein re-entrant
(turning back towards the body)

This is a small group of hoverflies which includes some of the biggest bee and wasp mimics.

▶ 4

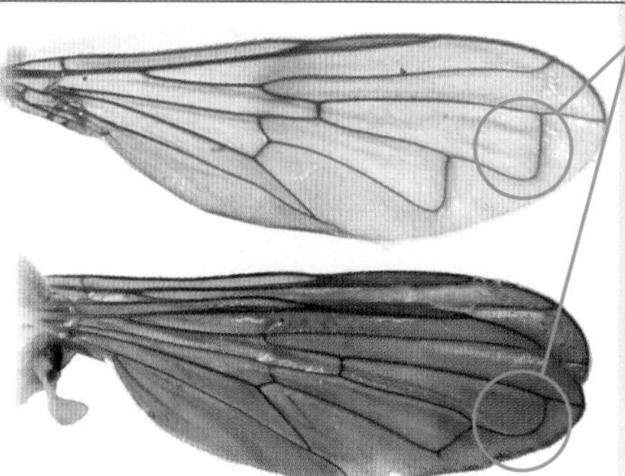

b) Upper outer cross-vein strongly upturned
(but not turning back towards the body)

Small narrow-bodied hoverflies

This feature is shared by two very similar genera *Neoascia* and *Sphegina*.

Chrysogastrini:
Neoascia *p.* 182
Sphegina *p.* 184

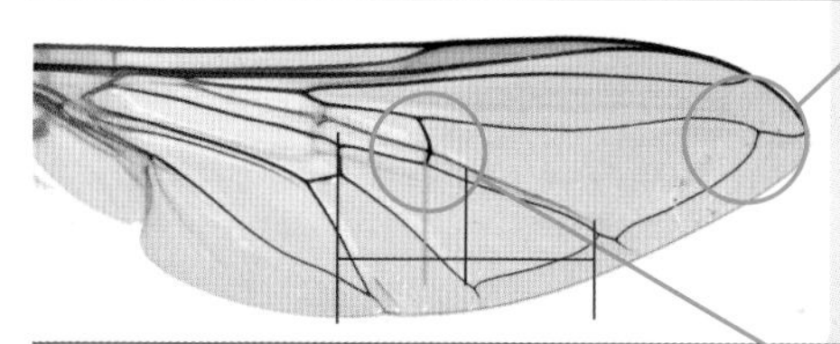

**c) 'Normal' wing –
Upper outer cross-vein neither re-entrant nor strongly upturned**

▶ 5

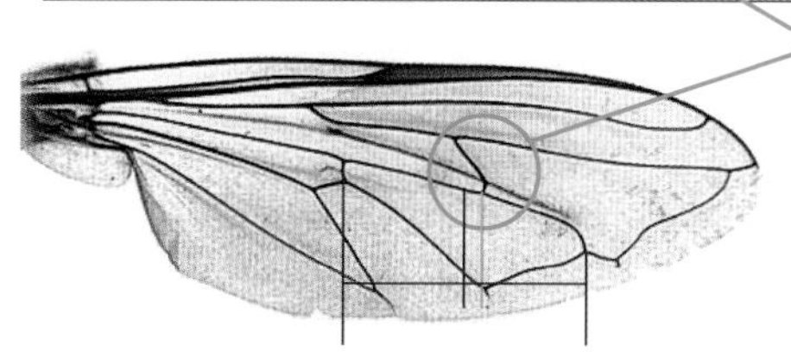

d) as c) but the inner cross-vein R-M meets the discal cell at a point at or beyond the middle of the cell

A heterogeneous group of bumblebee and hive-bee mimics together with more elongate species which resemble some sawflies.

Xylotini ***p.* 250**

Xylota sylvarum – a sawfly mimic

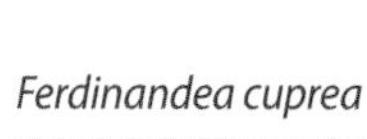

Ferdinandea cuprea

Beware – *Ferdinandea* [**Cheilosiini**] (*p.* 176) has similar wing venation, but looks very different, with a brassy, metallic abdomen; grey stripes running along the thorax; and a few bristles on the upper sides of the thorax.

4
from
3a

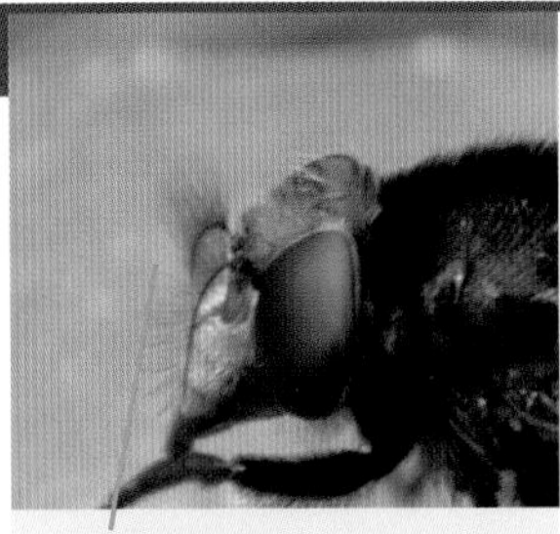

a) Aristae strongly plumose

Large bumblebee and wasp mimics.

Volucellini:
Volucella p. 244

b) Antennae long and forward-pointing ('porrect')

Medium-sized dumpy hoverflies with relatively short wings.

Microdontinae:
Microdon p. 272

c) Antennae 'normal' with a bare arista
Hind femur enlarged

Small, shiny or brassy hoverflies.

Merodontini:
Eumerus p. 220

4 Hoverflies with wings which have the upper outer cross-vein re-entrant

Microdontinae: *Microdon analis* p. 272

Merodontini: *Eumerus funeralis* p. 220

Bumblebee mimic
Volucellini: *Volucella bombylans* p. 236

Wasp mimic
Volucellini: *Volucella inanis* p. 248

5

from **3c**

a) Antennae 'porrect' with a terminal arista (which is white-tipped)

Callicerini: *Callicera* *p. 154*

The three species of *Callicera* are all rare and might be overlooked as solitary bees. Check – wing venation (presence or absence of a vena spuria) to ascertain whether it is a hoverfly or solitary bee.

b) Aristae not terminal; strongly plumose
Large hoverflies

Sericomyiini ***p. 240***

Bumblebee mimics	*Sericomyia suberbiens* *p. 240*
Wasp mimics	*S. lappona & silentis* *p. 244*

Beware – *Volucella* (*p. 244*) are also large with plumose aristae but with different wing venation (**3a**) and should already have keyed out at **4a**.

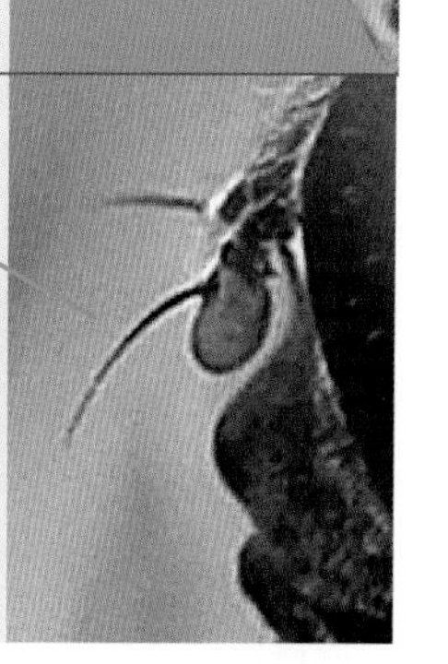

c) Aristae not terminal; bare or with weak pilosity

Antennae various shapes

Generally much smaller species

▶ **6**

5b Hoverflies with strongly plumose aristae and 'normal' wings

Bumblebee mimic
Sericomyiini: *Sericomyia superbiens* *p. 240*

Wasp mimic
Sericomyiini: *Sericomyia silentis* *p. 242*

6

from **5c**

a) Face: flat, with long drooping hairs

This is a group of mainly black hoverflies, sometimes with yellow spots on abdomen segment T2.

Pipizini ***p. 228***

Beware – *Psilota anthracina* (*p. 224*) may also key out here (see **2a**). It is a shining bluish-black hoverfly whose flat face has a strongly pointed mouth-edge.

b) Face strongly projected

Cheilosiini: *Rhingia p. 178*

Anasimyia lineata [Eristalini] (*p. 212*) has a similar but less extreme projection, although the strong wing loop distinguishes this species.

c) Face with some projection; Antennae unusual, 'half-moon' shaped and with a strongly thickened arista

Pelecocerini: *Pelecocera p. 226*

These are relatively small black-and-yellow hoverflies that are confined to the heathlands of southern England and to Scottish conifer woods.

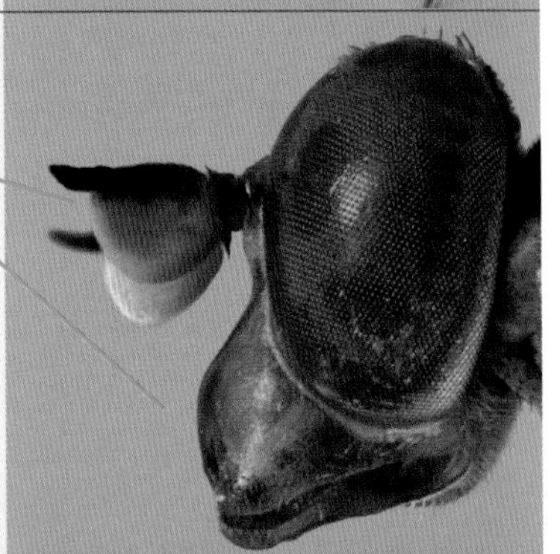

d) Face with some or no projection; Antennae 'normal' with a fine arista

▶ 7

NB the species illustrated is *Portevinia maculata*, note the zygoma which is diagnostic of the Cheilosiini (see **7a**).

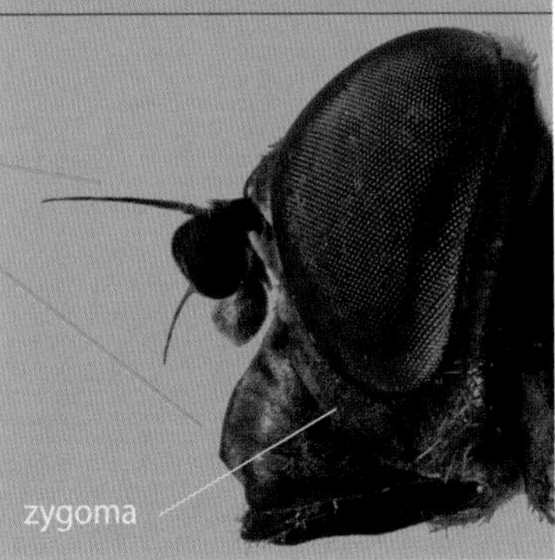

7a) Cheilosiini: *Cheilosia bergenstammi*

7b) Chrysogastrini: *Lejogaster metallina*

7 *from* **6d**

a) Face with a 'nose' above the mouth – giving the face a somewhat bulbous appearance

Zygoma present

Cheilosiini ***p. 158***

A heterogeneous group of hoverflies: most *Cheilosia* are black and unmarked, although some are very hairy, and are weak bee mimics; *Portevinia* (*p. 176*) has distinct markings; both *Rhingia* (see **6b**) and *Ferdinandea* (*p. 176*) are colourful and distinctive.

b) Face with no 'nose' – giving the face a strongly concave appearance

Zygoma absent

Beware – some species have a slight 'nose', but never as distinct as Cheilosiini (*p. 158*).

Chrysogastrini [part] ***p. 180***

A heterogeneous group of dark hoverflies; many of which have a metallic sheen under certain light conditions.

c) Face without projection or bristles:

Wholly brownish-orange
– can darken with age

Chrysogastrini: *Hammerschmidtia* *p. 196*

With a grey thorax and brownish-orange abdomen

Chrysogastrini: *Brachyopa* *p. 194*

Beware – many species in other fly families are similar in appearance – check the wing for a vena spuria to make sure it is a hoverfly!

Brachyopa scutellaris

8
from **1b**

a) Face ground colour black (although there may be paler dusting over the black ground colour)

Scutellum black

Bacchini ***p. 72***

Melanostoma mellinum

b) Face partially or wholly yellow

Scutellum either black or yellow

Beware – *Paragus* [Paragini] (*p. 92*) can have quite a dark face with only weak yellow areas towards the edges.

9
from **8b**

a) Tiny black flies with a distinctly narrowed 'waist' to the abdomen

Body length less than 5 mm

Face yellow with central black stripe.

Paragini: *Paragus p. 92*

b) Mainly colourful flies

Body length 5–12 mm

Both face and scutellum usually at least partially yellow, but a small number of species are darker *e.g. Leucozona laternaria* (*p. 112*) [black scutellum]. Some *Melangyna* (*pp. 128–133*) [very dark faces].

Syrphini ***p. 94***

Simplified guide to British hoverfly tribes

Showing the set of key features to look for when assigning a specimen.

The number of genera within each tribe and the number of species illustrated in the book [total number of species within the tribe] are also given.

<table>
<tr><th>SYRPHINAE:</th><th colspan="2">THORAX: front obscured; humeri bare:</th><th>Gen/Spp</th></tr>
<tr><td>BACCHINI
p. 72</td><td colspan="2">FACE: black
SCUTELLUM: black</td><td>4 genera
15 [30] spp.</td></tr>
<tr><td>PARAGINI
p. 92</td><td colspan="2">FACE: yellow, often obscure, but always with central black stripe
FORM: tiny (≤5mm) black</td><td>1 genus
1 [4] spp.</td></tr>
<tr><td>SYRPHINI
p. 94</td><td colspan="2">FACE: some yellow, some with a central black stripe, some black
FORM: larger (≥5–12mm); colourful
SCUTELLUM: often yellow, at least in part</td><td>18 genera
52 [84] spp.</td></tr>
<tr><th>ERISTALINAE:</th><th colspan="2">THORAX: front visible; humeri hairy</th><th>Gen/Spp</th></tr>
<tr><td>ERISTALINI
p. 198</td><td colspan="2">WING: strong wing loop R4+5 (cf. Merodontini: Merodon)
LEGS: black/yellow</td><td>8 genera
20 [28] spp.</td></tr>
<tr><td>VOLUCELLINI
p. 244</td><td colspan="2">WING: outer upper cross-vein re-entrant
ARISTA: plumose (cf. Merodontini: Eumerus, Microdontinae)</td><td>1 genus
5 [5] spp.</td></tr>
<tr><td>MERODONTINI
p. 220</td><td colspan="2">Heterogeneous tribe:
Psilota – WING: vena spuria absent
Merodon – WING: strong wing loop R4+5 (cf. Eristalini)
LEGS: all black
Eumerus – WING: outer upper cross-vein re-entrant
LEGS [hind femur]: enlarged</td><td>3 genera
6 [7] spp.</td></tr>
<tr><td>XYLOTINI
p. 250</td><td rowspan="7">wing: 'normal'
defined as one with no wing loop in R4+5 and the upper outer cross-vein neither re-entrant nor strongly upturned – see page 52 and 3c, page 57.</td><td>+ the inner cross-vein R-M meets the discal cell at a point at or beyond middle of the cell</td><td>10 genera
19 [20] spp.</td></tr>
<tr><td>CALLICERINI
p. 154</td><td>ANTENNAE: porrect
ARISTA: terminal with white tip</td><td>1 genus
3 [3] spp.</td></tr>
<tr><td>SERICOMYINI
p. 240</td><td>WING: outer upper cross-vein not re-entrant
ARISTA: plumose</td><td>2 genera
3 [3] spp.</td></tr>
<tr><td>PELECOCERINI
p. 226</td><td>ANTENNA: half-moon shaped
ARISTA: thickened</td><td>1 genus
2 [3] spp.</td></tr>
<tr><td>PIPIZINI
p. 228</td><td>FACE: flat with long drooping hairs
FORM: small–medium, predominantly black</td><td>5 genera
10 [20] spp.</td></tr>
<tr><td>CHEILOSIINI
p. 158</td><td>FACE: nose-like central prominence;
zygoma present</td><td>4 genera
20 [43] spp.</td></tr>
<tr><td>CHRYSOGASTRINI
p. 180</td><td>Heterogeneous tribe with no consistent features, predominantly small–medium-sized dark hoverflies with concave faces; some spp. metallic; a few spp. colourful - see Guide to Chrysogastrini for more information</td><td>10 genera
17 [29] spp.</td></tr>
<tr><td>MICRODONTINAE:
p. 272</td><td colspan="2">WING: outer upper cross-vein re-entrant
ANTENNAE: porrect (cf. Volucellini, Merodontini: Eumerus)</td><td>1 genus
4 [4] spp.</td></tr>
</table>

Identifying wasp and bee mimics

(see also *page 32*)

Two species that hoverflies mimic: *Andrena carantonica* (left) and *Osmia rufa* (right)

Honey-bee mimics

Species	Wing	Hind leg	Comments
Eristalis tenax *p.206*	loop in R4+5	bent and thickened with hairs on the tibia resembling a pollen basket	Good honey-bee mimic
Other *Eristalis* (except *E. intricaria*) *pp.200–205*	loop in R4+5	partly pale	Not very good bee mimics!
Mallota cimbiciformis *p.210*	loop in R4+5	swollen hind femora	Good honey-bee mimic
Criorhina asilica *p.260*			Good honey-bee mimic
Brachypalpus laphriformis *p.256*		swollen hind femora	Could be considered as mimics of solitary bees like *Osmia*
Chalcosyrphus eunotus *p.258*		swollen hind femora	

The Drone Fly, ***Eristalis tenax*** is a relatively good honey-bee mimic. Its hind legs have both the femur and tibia somewhat bent and the tibia has a patch of long, stiff, black hairs near the middle mimicking the pollen basket of a honey-bee. Other ***Eristalis*** (except ***E. intricaria*** which is a bumblebee mimic) are rather less convincing. ***E. pertinax*** (*p.200*) perhaps has the best claim to be a bee mimic and needs a careful look to distinguish from ***E. tenax***, but the yellow front and middle 'feet' are distinctive.

Another Eristaline (with the loop in R_{4+5}) which is a good honey-bee mimic is ***Mallota cimbiciformis***. This is a scarce species which is readily distinguised from ***E. tenax*** because it has an enlarged hind femur and lacks the characteristic bent hind legs and vertical bands of black hair down the eyes of ***E. tenax***.

Criorhina asilica can readily be recognised as a ***Criorhina*** because of the distinctive face-shape. It can require care to distinguish it from ***C. floccosa*** and ***C. berberina*** (*p.262*).

Brachypalpus laphriformis and ***Chalcosyrphus eunotus*** are both scarce species with swollen hind femora. See *page 257* for separation. They are sometimes described as mimics of solitary bees rather than of honey-bee.

Furry members of the genus ***Cheilosia*** (***C. albipila*** – *p. 168*, ***C. chrysocoma*** – *p. 174* and ***C. grossa*** – *p. 168*) are sometimes described as bee mimics but it is not always clear what sort of bee they are supposed to be mimicking!

Bumblebee mimics

Species	Wing	Scutellum	Hind leg	Arista	Comments
Sericomyia superbiens *p. 240*		pale	all black	**plumose**	All buff coloured, upland and western
Volucella bombylans *p. 244*	**re-entrant upper outer cross-vein**	varies depending on colour form	all black	**plumose**	Large, very good mimic
Eristalis intricaria *p. 206*	loop in R_{4+5}	dark	partially pale	has some longer hairs near base	sexually dimorphic; females have a white tail, males with a buff tail
Merodon equestris *p. 222*	loop in R_{4+5}	varies depending on colour form	all black with a flange on the femora	bare	Bewildering variety of colour forms – see *page 31*
Eriozona syrphoides *p. 112*		pale	partially pale	bare	Conifer forests; **yellow face unique among mimics**
Cheilosia illustrata *p. 162*		black	all black	bare	not a very convincing bee mimic! Face black
Pocota personata *p. 264*		black	all black	bare	Extremely good mimic; rare. **Very distinctive shape with the head looking too small for its body**
Criorhina berberina *p. 262*		pale in form *oxyacanthae* dark in typical form	all black	bare	**Distinctive face-shape that extends well below the bottom of the eye in profile**
Criorhina floccosa *p. 262*		pale	all black	bare	
Criorhina ranunculi *p. 260*		partially pale	all black	bare	

These ten species are generally considered as bumblebee mimics, but they range from really good examples like ***Pocota personata*** and ***Volucella bombylans*** to ***Cheilosia illustrata*** which cannot be regarded as very convincing! They mostly have black hind legs, except for ***Eriozona syrphoides*** and ***Eristalis intricaria***; these two species can be readily separated because ***E. intricaria*** has a loop in vein R_{4+5}.

Of the black-legged species, ***Merodon equestris*** has a loop in vein R_{4+5}, and a flange on the hind femur; it also has quite a distinctive shape.

The rare ***Pocota*** also has a very distinctive shape because its head looks too small for its body.

Volucella and ***Sericomyia*** are large species with plumose aristae, but ***Volucella*** has a re-entrant upper-outer-cross-vein.
The ***Criorhina*** species are distinctive because of the shape of their head in profile in which the face extends well below the eye margin. Of these ***C. berberina*** and ***C. floccosa*** can be quite tricky to tell apart and require considerable care to be sure of the identification.

Wasp mimics

Species	Wing	Hind leg	Arista	Comments
Volucella zonaria *p.248*	re-entrant upper outer cross-vein	dark	plumose	The two big, black-and-yellow striped *Volucella* are often regarded as hornet mimics. Care needed to tell them apart.
Volucella inanis *p. 248*	re-entrant upper outer cross-vein	dark	plumose	
Sericomyia silentis *p.242*		brown	plumose	Large, black and yellow, with plumose aristae
Chrysotoxum species *pp.98–103*		mainly yellow	long, porrect antennae with a bare arista	Reasonably good mimics of social wasps, especially *C. cautum*. The long, black antennae mimic those of social wasps quite successfully.
most *Syrphini* species *pp.94–153*		varies	bare	Black and yellow marked species are generally considered as wasp mimics, but are mostly rather unconvincing.

The black-and-yellow striped species of ***Volucella*** and ***Sericomyia*** are often described as hornet or wasp mimics. These are large hoverflies with plumose aristae. These two genera are easily told apart by their wing venation; ***Volucella*** has a re-entrant upper-outer cross-vein which ***Sericomyia*** lacks.
The two black-and-yellow striped *Volucella*, ***V. zonaria*** and ***V. inanis***, are not always easy to tell apart, especially in photographs taken from above. See *page 249* for their separation. (Note that there is a third black and yellow striped species, ***V. elegans***, in southern Europe which further complicates matters there).

Most black and yellow striped members of the tribe **Syrphini** are regarded as vaguely wasp-like in the popular press, but are generally rather unconvincing as wasp mimics! The exception is the genus ***Chrysotoxum*** which includes several rather good wasp mimics (especially ***C. cautum*** – *p. 100*).
In particular, the long, black antenae make a very good job of mimicking the "kneed" antennae of a social wasp in side view, especially when the 3rd segment is held drooped at an angle to the first two.
The rather arched shape of the abdomen also adds to the deception.

Wasp species that hoverflies mimic:
Hornet *Vespa crabro* (left) and Common Wasp *Vespula vulgaris* (right)

A guide to the most frequently photographed hoverflies

These plates show the 36 species that are most often photographed and aim to provide a visual guide to what you are most likely to see in urban gardens, parks, *etc.* Looking at pictures alone will not always get you to a firm identification but these plates should help to point you in the right direction if you are having trouble with the **Guide to the tribes** (see *page 55*). Some species/genera are very similar (*e.g. Epistrophe* and *Megasyrphus* are often mistaken for *Syrphus*). Colouration may be influenced by the temperature at which larvae develop (see *page 30*) and it is therefore not possible simply to rely on colour patterns. It also helps to check the distribution maps and flight time diagrams to rule out species that are unlikely to occur where and when a photo was taken! The section on **Photographing hoverflies** on *page 285* provides useful tips for obtaining good, identifiable images.

The smaller, grey images show the actual size of the hoverfly. The frequency ranking of each species is shown before its name (*e.g.* 1 to 36), but similar-looking species are shown together.

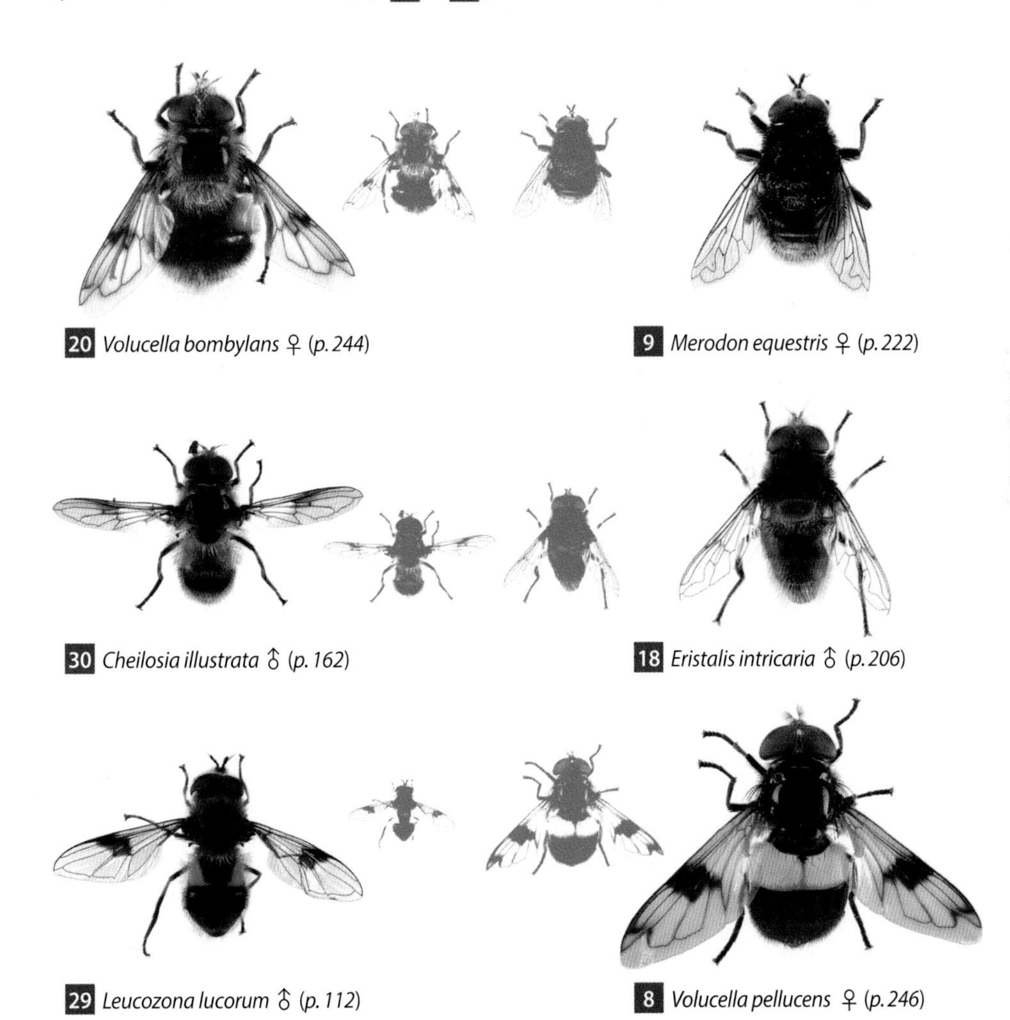

20 *Volucella bombylans* ♀ (*p. 244*)

9 *Merodon equestris* ♀ (*p. 222*)

30 *Cheilosia illustrata* ♂ (*p. 162*)

18 *Eristalis intricaria* ♂ (*p. 206*)

29 *Leucozona lucorum* ♂ (*p. 112*)

8 *Volucella pellucens* ♀ (*p. 246*)

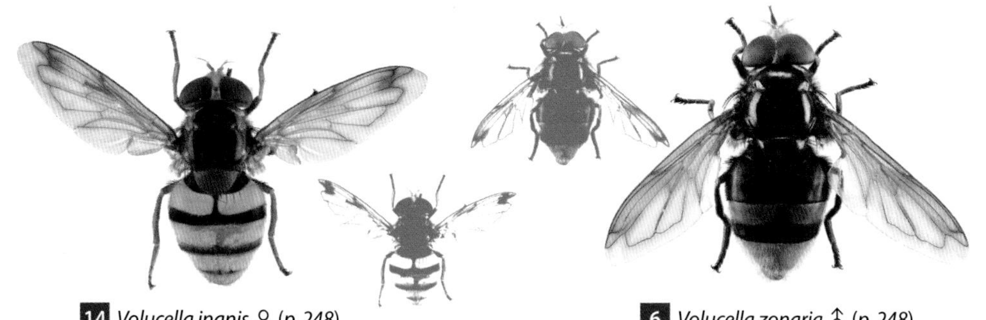

14 *Volucella inanis* ♀ (*p. 248*)

6 *Volucella zonaria* ♂ (*p. 248*)

16 *Sericomyia silentis* ♂ (*p. 242*)

28 *Chrysotoxum bicinctum* ♀ (*p. 98*)

1 *Episyrphus balteatus* ♂ (*p. 138*)

34 *Chrysotoxum festivum* ♂ (*p. 98*)

36 *Dasysyrphus albostriatus* ♀ (*p. 116*)

33 *Epistrophe grossulariae* ♀ (*p. 142*)

27 *Meliscaeva auricollis* ♀ (*p. 136*)

25 *Syrphus* sp. ♂ (*pp. 150–153*)

13 *Eupeodes luniger* ♂ (*p. 124*)

11 *Scaeva pyrastri* ♂ (*p. 122*)

15 *Eupeodes corollae* ♂ (*p. 126*)

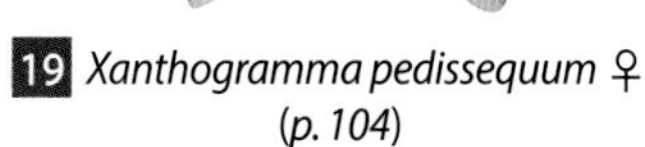

19 *Xanthogramma pedissequum* ♀ (*p. 104*)

7 *Melanostoma scalare* ♀ (*p. 76*)

35 *Leucozona glaucia* ♂ (*p. 114*)

23 *Epistrophe eligans* ♂ (*p. 144*)

22 *Platycheirus albimanus* ♀ ♂ (*p. 80*)

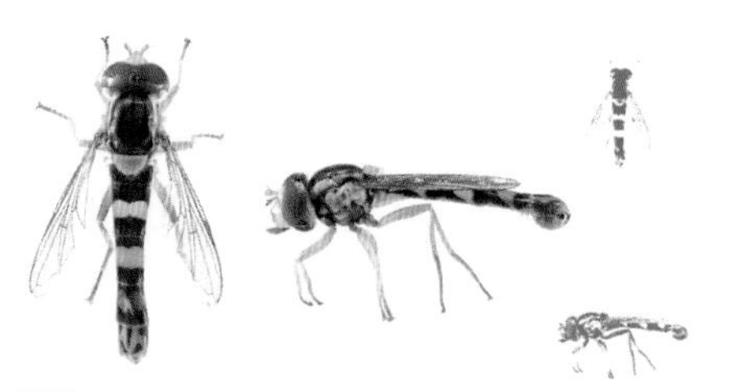

17 *Sphaerophoria scripta* ♂ (*p. 110*)

10 *Syritta pipiens* ♂ (*p. 266*)

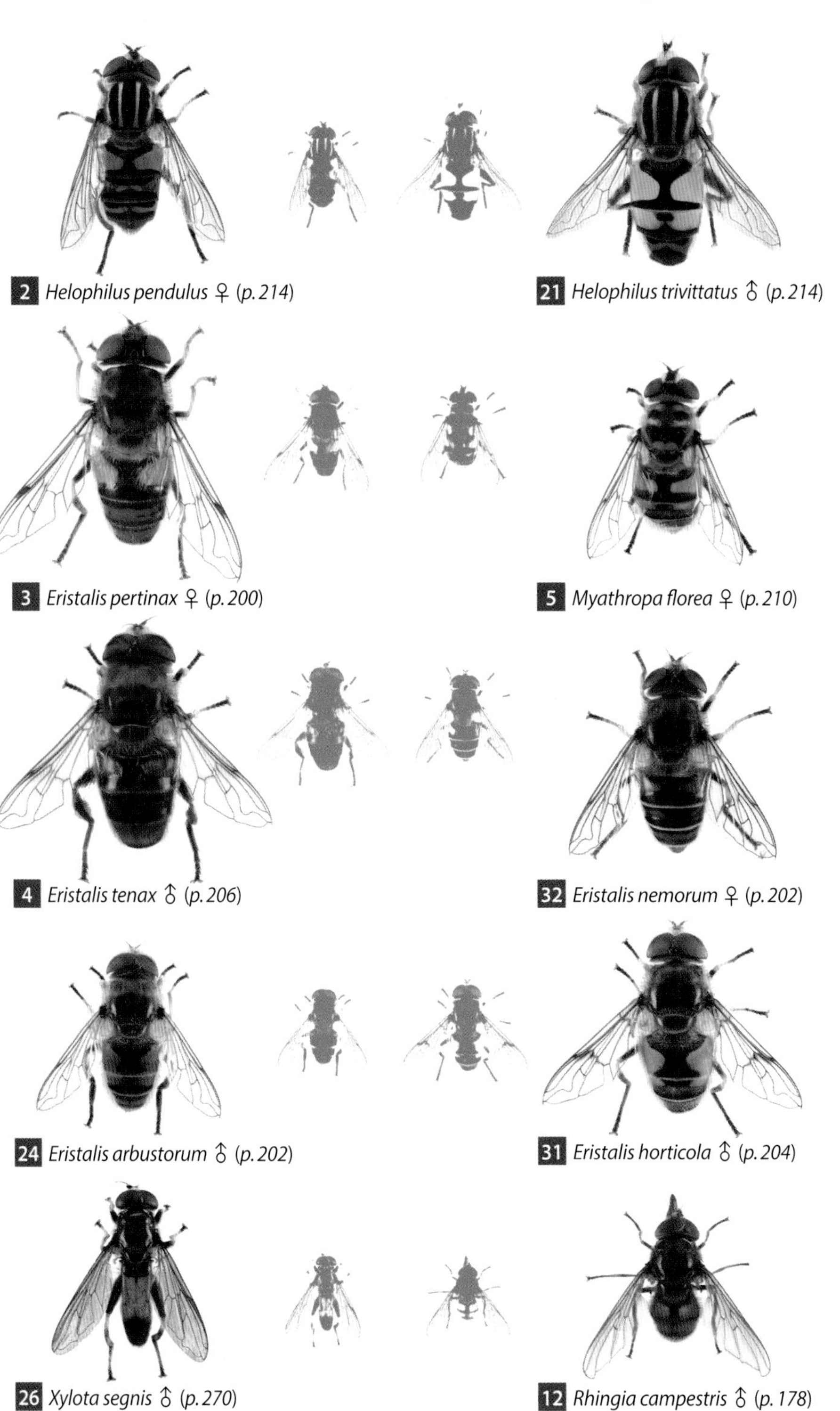

2 *Helophilus pendulus* ♀ (*p. 214*)

21 *Helophilus trivittatus* ♂ (*p. 214*)

3 *Eristalis pertinax* ♀ (*p. 200*)

5 *Myathropa florea* ♀ (*p. 210*)

4 *Eristalis tenax* ♂ (*p. 206*)

32 *Eristalis nemorum* ♀ (*p. 202*)

24 *Eristalis arbustorum* ♂ (*p. 202*)

31 *Eristalis horticola* ♂ (*p. 204*)

26 *Xylota segnis* ♂ (*p. 270*)

12 *Rhingia campestris* ♂ (*p. 178*)

The species accounts

ACCOUNT ORDER

The species accounts are arranged in taxonomic order by sub-family and tribe as follows: Syrphinae (3 tribes), Eristalinae (10 tribes) and Microdontinae (1 genus). The **Guide to the tribes** on *page 55* will direct the reader to the relevant page for that tribe (in some cases genus or species). Each tribe is prefaced by a guide to that tribe, which details the key identification features and presents a key to the genera within that tribe.

GENERA

Each genus is prefaced by a brief introduction to the main distinguishing features of that genus and a general statement about the larval biology. Genera that contain larger numbers of species may have a table of 'pointers' to assist in identification to species.

SPECIES

The order in which species within a genus are presented is based on the following criteria of priority: 1) closely-related and similar-looking species; 2) alphabetical order – **for readers wishing to find a particular genus, an index of genera is given on the inside back cover**.

SPECIES ACCOUNTS

Each account has information on identification, similar species and observation tips. Wing length is given, as text and graphically as a max-min bar. Icons for each species are designed to let the reader know how straightforward a species is to identify.

Can be identified in the field (with experience) without necessarily catching it or getting very close.

Can be identified in the field, but needs close and careful examination. This requires at least a very close approach and often requires temporary capture for examination using a hand-lens.

Requires examination at high magnification and in good lighting – this normally means a dead specimen under a microscope.

In tables, species illustrated are in ***bold italic*** species not illustrated in *italic*. Vagrant species are in *purple text*. The distribution maps for each species are colour coded for frequency of records. Beneath each map is a chart showing the flight period. In addition, an indication of frequency and any threat status are given.

BAP: Near Threatened

Frequent

J F M A M J J A S O N D

BAP (see *page 296*) and **threat status**. Rare and threatened species that are not illustrated but feature in a table are in *red italics* with an indicator of status as follows: Nationally Scarce [NS]; Near Threatened [NT]; Vulnerable [VU]; Endangered [EN] or Critically Endangered [CR]. Data Deficient species are coded [DD].

Frequency: the number of 10km squares of the Ordnance Survey National Grid in Great Britain the species has been recorded from.

Widespread	> 500 squares
Frequent	121–500 squares
Local	41–120 squares
Scarce	16–40 squares
Rare	<15 squares

Distribution: as recorded by the Hoverfly Recording Scheme as of May 2011.

most records

some records

few records

Flight Period:
The frequency of records from each week.

Peak records — No records

MAIN IMAGES – Bearing in mind the challenges of photos and perspective, an attempt has been made to present the species to scale – this approximate scale is given for each species.

Guide to Bacchini

Although there are only four genera in the Bacchini, one of them (*Platycheirus*) is our second largest and its species can be difficult to identify. **The tribe is characterised by the absence of hairs on the humeri, a somewhat concave head which closely fits the thorax, a black face, a black scutellum** and markings on the abdomen which are usually yellow or orange, but can consist of silvery-grey or bronzy-coloured dust spots. Occasional all black 'melanics' are found and these may cause confusion with the **Cheilosiini** (*p. 158*) or **Chrysogastrini** (*p. 180*) which also have a black face but are generally broader-bodied.

1

a) Small; abdomen extremely elongate, narrow and waisted

Distinctive – apart from the genera *Sphegina* and *Neoascia* [both Chyrsogastrini (*p. 182,184*)] which are somewhat smaller and, whilst they have narrow, 'waisted' abdomens, are not so elongate.

Baccha elongata p. 74

Baccha elongata

b) Large; oval-bodied; abdomen broad; scutellum black; distinctive markings on abdomen

Xanthandrus comtus p. 90

c) Narrow-bodied; abdomen not waisted or broad

2

Xanthandrus comtus

2
from **1c**

a) FEMALES: eyes separated at top of head

▶ 3

b) MALES: eyes touch at top of head

▶ 4

3
from **2a**

a) ♀ – Abdomen with distinctive triangular markings on abdomen segments T3 + T4

Melanostoma p. 76

b) ♀ – Abdominal markings not triangular

Platycheirus pp. 78–91

4
from **2b**

a) ♂ – Front legs modified in some way; or the abdomen with distinctive markings

Note: *Platycheirus granditarsus* (*p. 88*) and *P. rosarum* (*p. 90*) have abdominal markings that do not conform to the general pattern.

See note below.

Platycheirus pp. 78–91

b) ♂ – Front legs not modified

Melanostoma p. 76

See note below.

4a) *Platycheirus* front leg modifications

Most *Platycheirus* males have modified front legs, the nature of which and the location of hairs are useful in identification. However, both *P. rosarum* (*p. 90*) and *P. ambiguus* (*p. 80*) have front legs without any tarsal modification; and could lead to confusion with *Melanostoma* (*p. 76*).

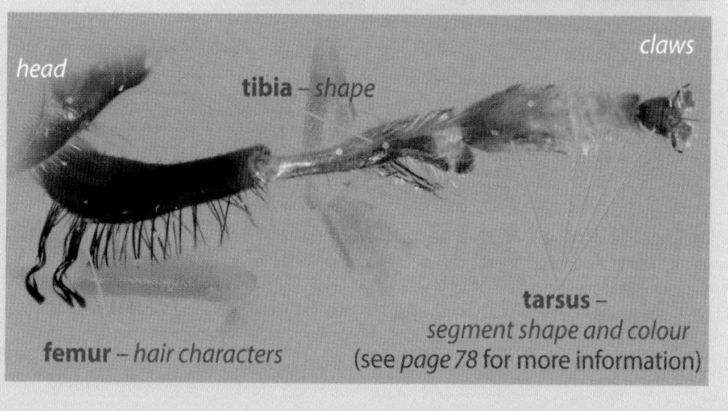

Baccha

1 British species (illustrated)

Small, dark and with a very elongate, wasp-waisted abdomen.
The larvae feed on a variety of aphids, especially Nettle Aphid and Bramble Aphid.

Baccha elongata

Widespread

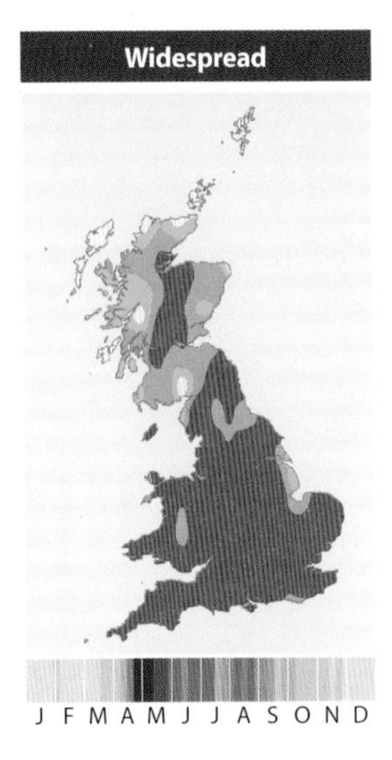

Wing length: 4·0–8·25 mm

Identification: The very long, slender, wasp-waisted shape should make this small hoverfly unmistakable.

Similar species: Only *Neoascia* and *Sphegina* (Chrysogastrini: see *page 180*) are anything like *Baccha* in shape. *B. elongata* has a longer and more slender abdomen and lacks the swollen hind femora found in *Neoascia* and *Sphegina*. In addition, the hairy humeri and distinctive wing venation of these two genera should ensure there is no confusion. In these genera, the upper outer cross-veins are sharply upturned and join R_{4+5} at right-angles (illustrated in *Guide to Tribes*: 3b on *page 57*), whereas in *B. elongata* it joins R_{4+5} at a shallow angle.

Observation tips: Dappled shade in woodland, along hedgerows and similar places where it manoeuvres low down amongst the vegetation. Does not often visit flowers, but does like to bask on sunlit leaves. Although widespread and common, it is inconspicuous and often overlooked.

A male *B. elongata* hovering.

▲ ♀ – *Baccha elongata* × 10 – ♂ ▼

Melanostoma

3 British species (2 illustrated)

Small, yellow-and-black hoverflies with a completely black face and scutellum that are best found by sweeping in grassy places. **Females are readily identified by their distinctive abdominal markings. Males are most likely to be confused with *Platycheirus* (*pp. 78–91*), but do not have the modified front legs prevalent in that genus.** The larvae feed on a variety of aphids amongst leaf-litter and the ground layer. Adults often visit ostensibly wind-pollinated flowers, such as plantains, grasses and sedges, to feed on pollen. Dead hoverflies found hanging below grass and flower heads are often *Melanostoma* that have been killed by the fungus *Entomophthora muscae*.

Melanostoma mellinum

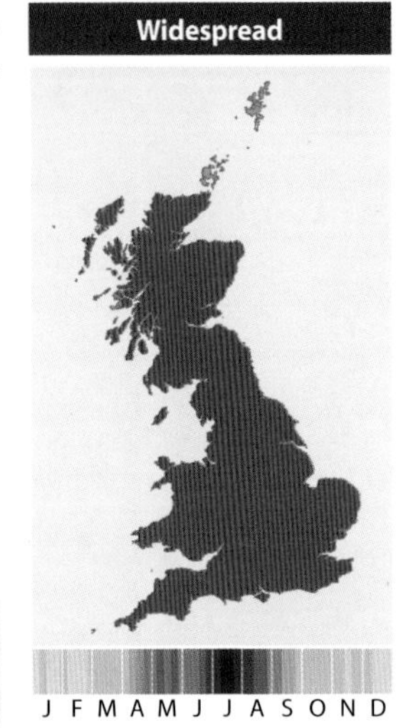

Wing length: 4·75–7 mm

Identification: Separation from *M. scalare* requires careful examination. Females have very narrow dust spots on the mostly shining black frons. Males have a relatively short abdomen in which segments T2 and T3 are no longer than they are wide. In the uplands it is necessary to check carefully for the Nationally Scarce *M. dubium*.

Observation tips: A widespread and abundant grassland species which can be numerous in the uplands, especially around wet flushes on moorland.

Similar species: All *Melanostoma* species are very similar. They could be confused with some *Platycheirus* (see *pp. 78–91*) and possibly *Pelecocera* (*p. 226*), but any confusion is readily resolved by careful examination.

Melanostoma scalare

Wing length: 5·5–8 mm

Identification: Separation from *M. mellinum* requires careful examination. Females have broad dust spots, extending most of the way across the frons. Males have an elongate abdomen in which the segments T2 and T3 are much longer than they are wide, giving them a distinctive appearance that is relatively easy to pick out with experience.

Observation tips: A widespread and abundant grassland species, but less frequent in the uplands. Often found in less open situations than *M. mellinum* such as woodland rides and scrubby grassland. It has a very long flight season, and is more prominent in Spring than *M. mellinum*.

Melanostoma species not otherwise covered:

M. dubium – Work by the Malloch Society suggested that this may be the extreme end of an altitudinal cline in the very variable *M. mellinum*. Whether it is a good species or simply a variety, it is restricted to high altitude - mainly in Scotland.

Platycheirus — 25 British species (11 illustrated)

Small, long-and-narrow hoverflies with yellow, or occasionally silvery, abdominal markings, black faces and a black scutellum. The males of many species have ornamented front legs with broad, flattened tarsi. *Platycheirus* is the second largest genus of British hoverflies. Some species are readily identifiable but there are many that are difficult, especially amongst females, and reference to a specialist key (*Stubbs & Falk 2002*) is required to attempt any identification. They can be abundant, but as the species look very similar in the field, it is necessary to retain plenty of specimens if you want to maximise your chances of finding all the species that occur at a site. The larvae feed on aphids and although some are specific to a few prey species, most seem to be general predators of leaf-litter and ground-layer aphids.

Platycheirus identification features.

Notes on the identification of British *Platycheirus* species, including species not otherwise covered:

	MALES	FEMALES
***ALBIMANUS* group** (*p. 80*) **– abdomen markings usually silvery or bronzy**		
P. ambiguus	LEGS: unmodified front tibia + tarsus with curled bristle at tip of femur	ABDOMEN: fairly distinctive because the markings are bands (not spots)
P. albimanus	These three species are difficult to tell apart, ID relies on details of the shape and hairing of the front tibia/tarsus	Difficult to distinguish using shape/markings of abdomen segments and dusting of the frons
P. discimanus [NS]		
P. sticticus [NS]		
***MANICATUS* group** (*p. 82*) **– abdomen markings yellow to orange**		
P. manicatus	THORAX: dull dusted	ABDOMEN: *differences in shape of markings*
P. tarsalis	THORAX: shining	
P. melanopsis [NT]	*Occurs at high altitude, mainly in Scotland*	
***PELTATUS* GROUP** (*p. 84*) **– abdomen markings yellow to orange**		
P. peltatus	LEGS: *distinguished by details of shape of front tarsi and hairing on middle tibia*	Not always possible to separate; based on shape of abdomen segments
P. nielseni		
P. amplus [NT]		Female unknown
***SCUTATUS* group** (*p. 84*) **– abdomen markings yellow to orange**		
P. scutatus	LEGS: *distinguished by details of hairs on middle tibia*	Females cannot be separated at present
P. aurolateralis		
P. splendidus		
***CLYPEATUS* group** (*p. 86*) **– abdomen markings yellow to orange**		
ABDOMEN: **yellow markings on T4 don't reach hind margin; T5 & T6 with little or no yellow**		
P. angustatus	Fairly distinctive narrow species with shiny sides to thorax in both sexes	
P. podagratus	LEGS: front tibia abruptly broadens near tip	Females difficult to distinguish using shape and markings on abdomen segments
P. clypeatus	LEGS: *distinguished using details of the markings on the front tarsi*	
P. europaeus		
P. occultus		
P. ramsarensis		
ABDOMEN: **yellow markings on T4 close to hind margin; T5 & T6 with extensive yellow**		
P. fulviventris	LEGS: male front tibia distinctive: broadens from near base	Females very difficult to distinguish – it may not be possible to separate reliably females of *P. immarginatus* and *P. perpallidus*
P. immarginatus [NS]	LEGS: *distinguished by details of the hairs on the front tibia and the shape of the front tarsi*	
P. perpallidus [NS]		
P. scambus		
Subgenus *PYROPHAENA* (*pp. 88, 90*) **– readily identified by characteristic abdomen patterns**		
P. granditarsus	LEGS: very strongly modified	ABDOMEN: characteristic pattern, but very variable
P. rosarum	LEGS: unmodified	ABDOMEN: characteristic pattern

Platycheirus ambiguus

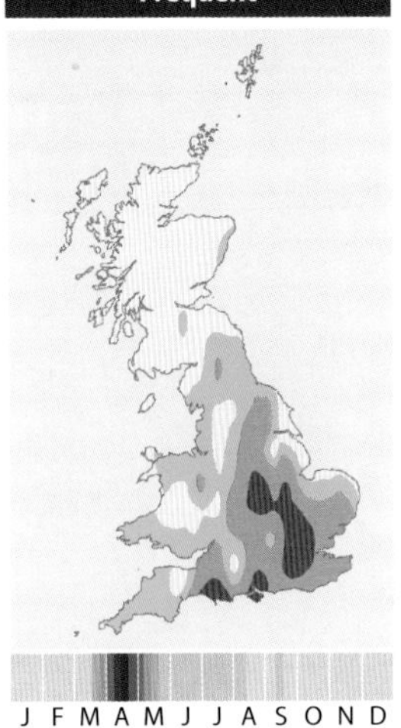

Wing length: 4·5–7 mm

Identification: Easily overlooked amongst others of the genus. Females are similar to *P. albimanus*, but have grey bands (rather than spots) on abdomen segments T3 and T4 which are sometimes quite faint. The front leg of the male has a very distinctive, curious, curled bristle near the apex of the femur (although this is only visible on close examination).

Observation tips: An early Spring species which occurs along woodland edge and hedgerows. They are most frequently found visiting Blackthorn blossom and males often hover close by, sometimes in small swarms. Most frequent in south east England.

Similar species: *Platycheirus albimanus, P. ambiguus* and the Nationally Scarce and rarely reported *P. discimanus* and *P. sticticus* (neither illustrated) are all very similar and all may occur together – see table on *page 79*.

Platycheirus albimanus

Wing length: 5–8 mm

Identification: The females are readily identifiable because the abdomen is black with grey spots and the legs are extensively yellow. Males have dark metallic bronze or silvery-grey markings on the abdomen and very distinctive front legs with a clump of tangled hairs near the base of the front femora. The front legs of male *P. albimanus* are similar to those of *P. aurolateralis, P. scutatus* (illustrated on *page 85*) and *P. splendidus*. All have similar hair tufts on the back of the front femur, but the 2nd to 4th tarsal segments of the SCUTATUS group are very narrow. Confusion with other genera is unlikely, but occasionally reports of *Portevinia maculata* (*p. 176*) have proved to be *P. albimanus*.

Observation tips: Widespread, abundant and found throughout the year, but perhaps most numerous in the Spring. It occurs amongst low foliage, such as Bramble and nettle patches, and at a very wide range of low-growing flowers. Common in gardens.

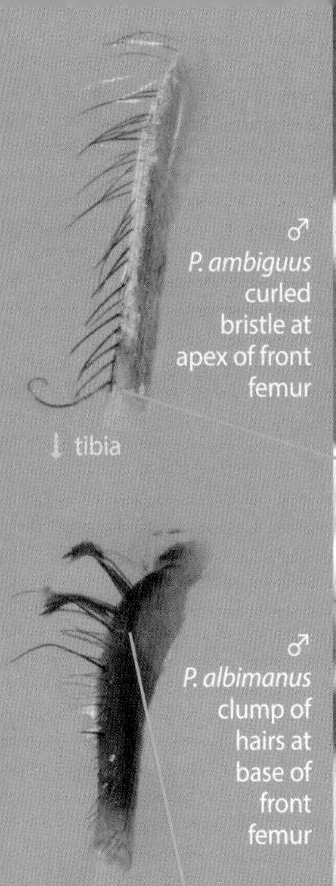

Platycheirus ambiguus ♂ ×**8**

♂ – *Platycheirus albimanus* ×**8** – ♀

Platycheirus manicatus

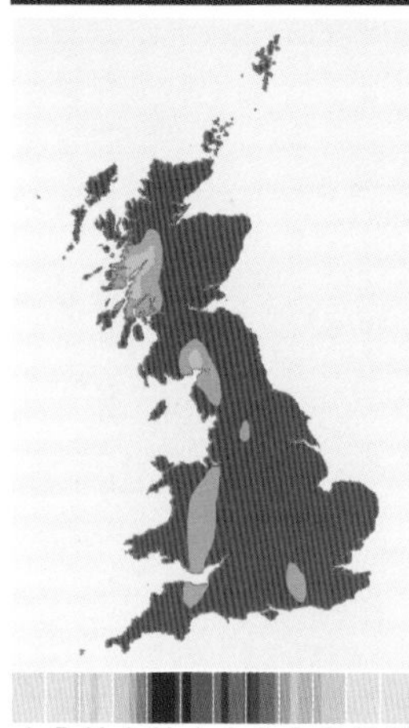

Wing length: 6·75–9 mm

Identification: There are three species with similar facial profiles in which the mouth margin is strongly extended forwards. Both sexes of *P. manicatus* have the thorax and head extensively dusted giving them a dull, slight bronzy appearance. In *P. tarsalis* and *P. melanopsis* the thorax is shiny. The male has front legs with tarsal segments 3–5 all the same mid-grey brown (or darker) colour (tarsal segment 5 paler and contrasting with darker segments 3 and 4 in *P. tarsalis*).

Similar species: *Platycheirus tarsalis* and the Nationally Scarce *P. melanopsis* (not illustrated) – see table on *page 79*.

Observation tips: A dry grassland species which often occurs on calcareous soils, coastal grasslands and moorlands. It can be one of the commonest species in exposed and open habitats in the north and west of Scotland and the Northern Isles.

Platycheirus tarsalis

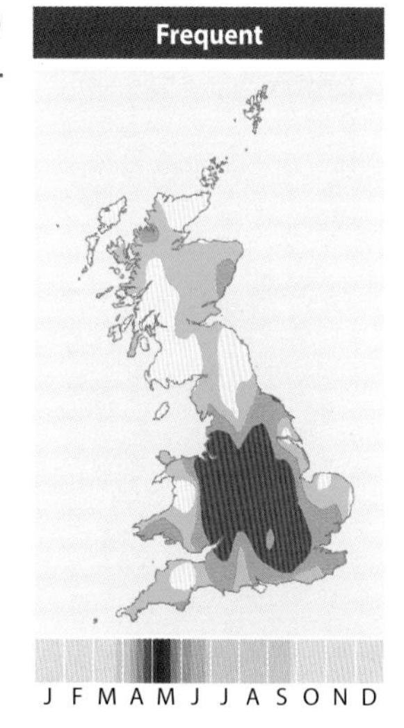

Wing length: 7·5–8·75 mm

Identification: This is one of three *Platycheirus* species with extended facial profiles, the others being *P. manicatus* and *P. melanopsis*. It is most likely to be confused with *P. manicatus*, but has a shiny black thorax (rather than dull and dusted). The male has front legs with tarsal segments 3–5 that tend to be quite dark with the 5th segment paler and contrasting (tarsal segments 3–5 all the same colour in *P. manicatus*).

Similar species: *Platycheirus manicatus* and the Nationally Scarce *P. melanopsis* (not illustrated) – see table on *page 79*.

Observation tips: A woodland species, usually found in open spaces such as rides and clearings. Although widespread, it is most abundant in the Midlands and the Welsh border counties. It mainly flies in May and June.

Platycheirus manicatus ♂ ×**8**

Platycheirus tarsalis ♂ ×**8**

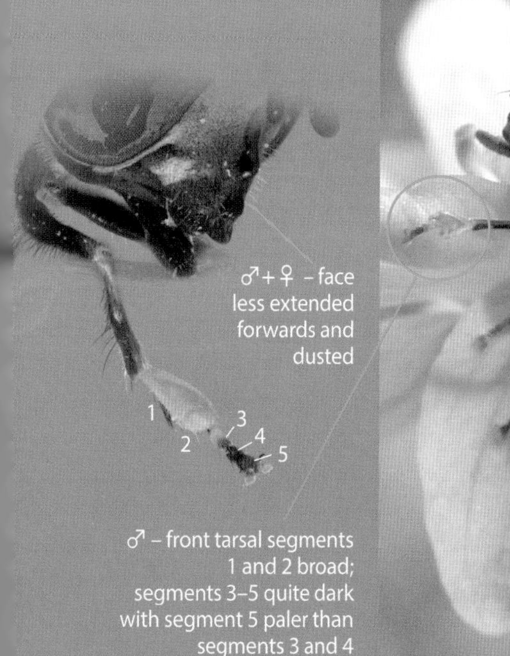

Platycheirus peltatus

Wing length: 7–9 mm

Identification: A comparatively large member of the genus with a relatively broad abdomen. It can only be confused with *P. amplus* and *P. nielseni*. Males have a distinctively shaped front metatarsus and characteristic hair patterns on the underside of the mid-tibia (careful examination is essential). Females are much trickier (and not always possible) to separate, especially from *P. nielseni*.

Similar species: *Platycheirus amplus* and *P. nielseni* (neither illustrated) are very similar– see table on *page 79*.
The *P. scutatus* group and possibly *Meliscaeva auricollis* (*p. 136*) may be confused in photographs.

Observation tips: A widespread species in lowland Britain, especially in damp woodland rides and ditches. The similar *P. nielseni* is a more northern and upland species than *P. peltatus*, but the distribution of the two species overlaps substantially in lowland areas in the north.

THE *SCUTATUS* GROUP

Platycheirus scutatus

Wing length: 5–7·5 mm

Identification: A small, narrow *Platycheirus*. Both sexes have clear yellow markings on the abdomen and antenna segment 3 yellow underneath. The front legs of the male are very distinctive, with tufts of tangled hairs both at the base of the femur and in the middle of the tibia, and with the metatarsus much longer than any of the other segments. Females cannot be separated from those of *P. aurolateralis* and *P. splendidus*.

Similar species: *Platycheirus aurolateralis* and *P. splendidus* (neither illustrated – see table on *page 79*). It is also possible that confusion may occur with the PELTATUS group and also perhaps with *Meliscaeva auricollis* (*p. 136*) in some photographs.

Observation tips: A widespread species, often found in association with *P. albimanus* in woodland rides, along the margins of grasslands at the interface with scrub and woodland, and also in gardens. It has a long flight period.

♂ – *Platycheirus peltatus* × **8** – ♀

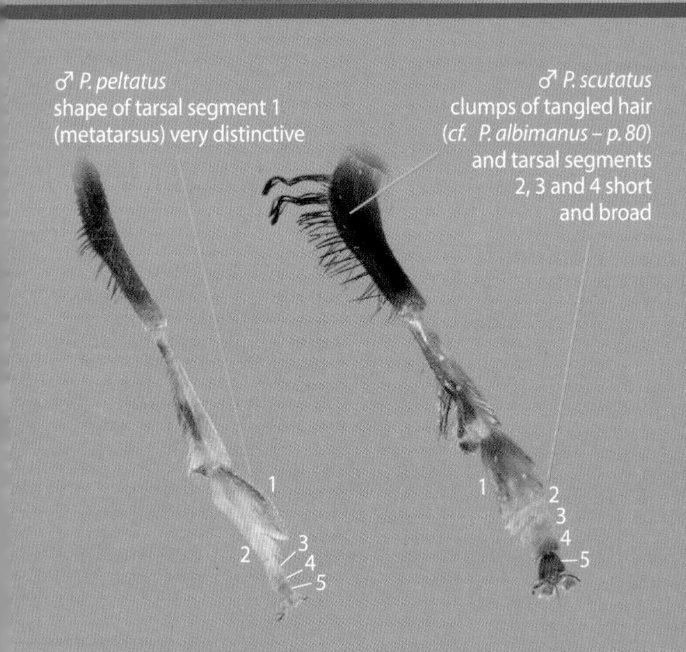

Platycheirus scutatus ♂ × **8**

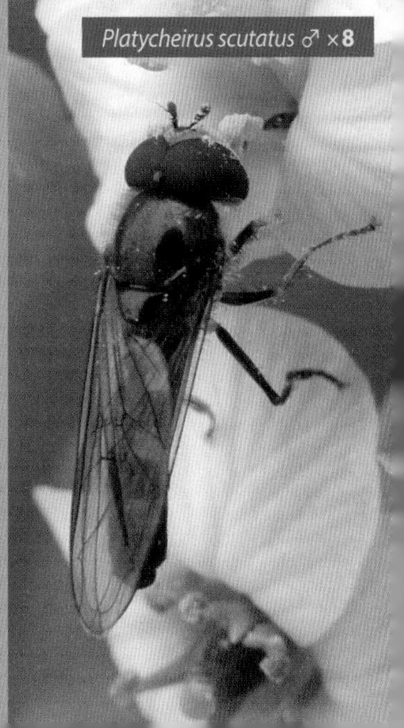

Platycheirus angustatus

Widespread

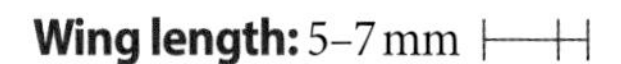

Wing length: 5–7 mm

Identification: A fairly distinctive, narrow species with shiny sides to the thorax in both sexes. This is a relatively straightforward species to identify under high magnification, but cannot be identified with certainty in the field. Males have distinctive 'lightning-shaped' pits on the undersides of the front metatarsi that can only be confused with *P. ramsarensis* and *P. europaeus*, which differ in details in the shape and arrangements of the pits. In both sexes, the sides of the thorax have large, shiny black patches.

Similar species: *Melanostoma mellinum* (male) (*p. 76*), *Platycheirus clypeatus*, *P. podagratus*, *P. europaeus*, *P. occultus*, *P. perpallidus* and *P. ramsarensis* (none illustrated) – see table on *page 79*.

Observation tips: A widespread and fairly common species which occurs mainly in damper situations in woodlands, grasslands and wetlands. It is frequently found at plantain and sedge flowers, as are similar species such as *P. clypeatus*.

Platycheirus clypeatus 

Widespread

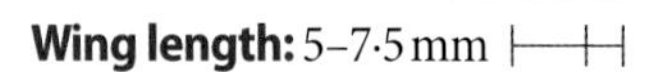

Wing length: 5–7·5 mm

Identification: This is one of a group of very similar species with completely black antennae which are difficult to identify. Males have distinctive elongate pits on the underside of their front metatarsi, but can be confused with *P. occultus* and so very careful examination is required. Females are much harder to separate from other very similar species in the genus: *P. angustatus*, *P. immarginatus*, *P. occultus*, and *P. scambus* which differ subtly in the shape and markings of the abdominal segments.

Similar species: *Melanostoma mellinum* (male) (*p. 76*), *Platycheirus europaeus*, *P. immarginatus*, *P. occultus*, *P. perpallidus*, *P. podagratus*, *P. ramsarensis* and *P. scambus* (none illustrated) – see table on *page 79*.

Observation tips: A common and widespread grassland species which is often found at plantain, grass and sedge flowers. It occurs from April onwards but becomes more frequent by mid-Summer.

Platycheirus angustatus ♀ ×**8**

♂ – *Platycheirus clypeatus* × **8** – ♀

Platycheirus fulviventris

Wing length: 6–7·5 mm

Identification: One of four *Platycheirus* species where the orange markings on the abdomen are very extensive (including abdomen segments T5 and T6) giving a rather colourful appearance, and the antennae are completely black. The front legs of the male are distinctive with the front tibia very broad for most of its length. Females have sharply defined dust markings on the frons (fuzzy-edged in the other three species).

Similar species: *Platycheirus immarginatus*, *P. perpallidus* and *P. scambus* (none illustrated) – see table on *page 79*.

Observation tips: A wetland species which occurs amongst vegetation fringing ponds, grazing marsh ditches and river banks. Most abundant in the south, scarcer further north.

PLATYCHEIRUS WITH CHARACTERISTIC ABDOMEN MARKINGS

Platycheirus granditarsus

Wing length: 5·25–8·5 mm

Identification: A readily recognised species in which both sexes have large orange-red markings on the abdomen. Occasionally, the abdominal markings in the females are more limited, making it more difficult to recognise. The male has both the front and middle legs modified with black, peg-like extensions to the tarsi.

Similar species: Possibly superficially confused with *Xylota segnis* (*p. 270*), which also has orange markings on the abdomen.

Observation tips: A widespread and often abundant wetland hoverfly which occurs in wet meadows and along the edges of water bodies such as ponds and ditches.

♂ – *Platycheirus fulviventris* × **8** – ♀

♂ – *Platycheirus granditarsus* × **8** – ♀

Platycheirus rosarum

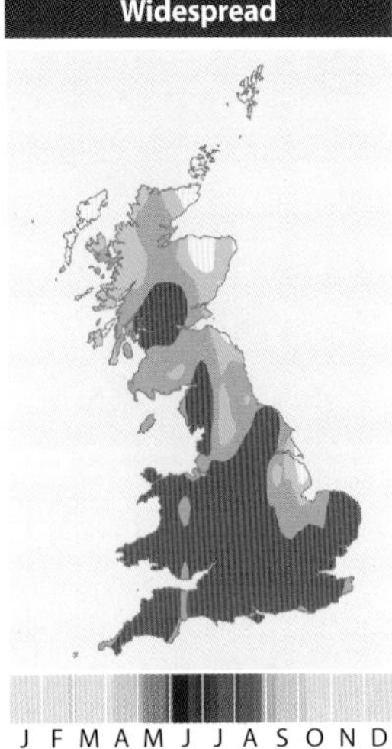

Wing length: 5·25–7·75 mm

Identification: Not immediately recognisable as a *Platycheirus* because there are no obvious modifications to the legs of the male. However, both sexes are readily recognised by the unusual shape and markings of the abdomen – which widens steadily towards the tip, and has a pair of pale yellow spots on abdomen segment T3 on an otherwise black background.

Similar species: None.

Observation tips: A widespread, but local species of wet meadows and along the edges of water bodies such as ponds and ditches. Often abundant where it occurs.

Xanthandrus (1 British species illustrated)

A reasonably large, robust hoverfly that has yellow-and-black abdominal markings and a completely black face and scutellum. The shape of the markings on the abdomen are quite distinctive. Larvae feed on colonial micro-moth caterpillars on a variety of trees and bushes.

Xanthandrus comtus

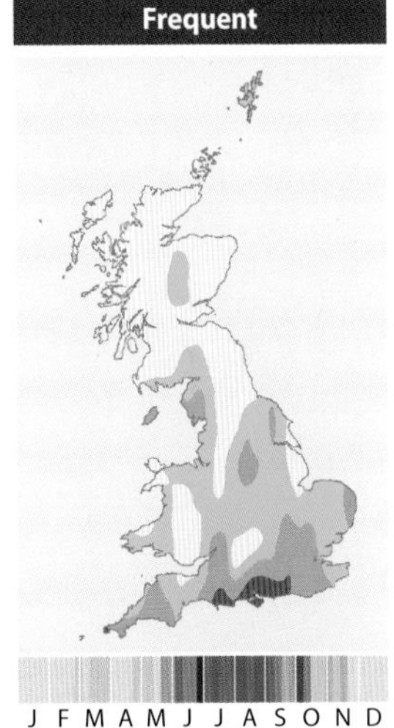

Wing length: 8·75–11·5 mm

Identification: Abdomen segment T2 has round spots and the segments T3 and T4 have semi-circular spots which may join in the middle. Males are considerably more colourful than females, which have darker and greyer markings on the abdomen.

Similar species: None.

Observation tips: Widespread but scarce and more frequent along the south coast. Regarded as a migratory species and there is no doubt that it occurs in very variable numbers each year. Often in scrubby localities with plenty of umbellifer flowers which the adults visit.

♂ – *Platycheirus rosarum* × **8** – ♀

Xanthandrus comtus ♂ × **8**

Xanthandrus comtus ♀ × **8**

Guide to Paragini

This tribe is represented by a single genus of tiny (usually less than 5 mm long) black hoverflies with a face that is yellow, at least at the sides (careful examination required as this can be a subtle feature). **They can be told from other similar looking species by the bare humeri, which are obscured by the head.**

Paragus

4 British species (1 illustrated)

Tiny hoverflies that fly close to the ground and visit low-growing flowers, especially yellow composites and Tormentil. They are usually black with few markings, but the legs are extensively yellow. **Careful microscopic examination is needed to ensure a correct identification.** The larvae feed on aphids.

Paragus haemorrhous

Widespread

Wing length: 3·5–4 mm

Identification: A tiny black hoverfly that has a slightly constricted abdomen. The face is partially yellow and the antennae are somwhat elongated. It needs to be examined in detail under the microscope to separate it from others in the genus.

Similar species: Other *Paragus*, but confusion most likely with *P. constrictus* (which is only known from Ireland within the British Isles) and the Near Threatened *P. tibialis* (neither illustrated), only the males of which can be identified by microscopic examination of the genitalia.

Observation tips: This is a grassland hoverfly which occurs in short swards in dry, sunny situations, often in the vicinity of cat's-ears. Adults rest on, or hover just above, patches of bare sand or earth along paths and banks.

Paragus face showing the partially yellow face and the elongate antennae.

Paragus haemorrhous ♂ ×**15**

Paragus species not otherwise covered:

P. albifrons [CR] – A species that has declined greatly: the few recent records are from the Thames Marshes, generally on shingle or dry grassland close to the coast.

P. constrictus – Only recorded from limestone pavements in Ireland.

P. tibialis [NT] – Modern records are restricted to heathlands in Hampshire, Surrey and Dorset.

Guide to Syrphini

A large tribe composed of 18 genera including nearly one third of our species and some of the most colourful and familiar. Most are moderate to large in size, but individuals of the same species can be very variable in size depending on seasonal fluctuations in the abundance of prey items for their larvae. **The absence of hairs from the humeri, together with the concave head shape (which fits snugly over the front of the thorax), and the presence of yellow markings on the face separate this tribe from all others apart from the Paragini – which, in Britain, are all tiny, black hoverflies –** (*p. 92*). Most have an abdominal pattern of black-and-yellow stripes or spots and most have a yellow scutellum, but there are exceptions such as some *Leucozona*. The patterns on the abdomen can be extremely helpful in separating them, but some patterns occur in several genera whose separation is then largely reliant upon microscopic characters. However, it should be remembered that members of other tribes have some common features – such as black-and-yellow abdominal patterns or yellow faces but these all have visible, hairy humeri.

1

a) Sides of the thorax with distinct, sharply-defined yellow markings; lacking dusting and only weakly hairy

Beware – it is the markings on the side of the thorax that are meant here, not the stripes running along the sides of the top surface of the thorax.

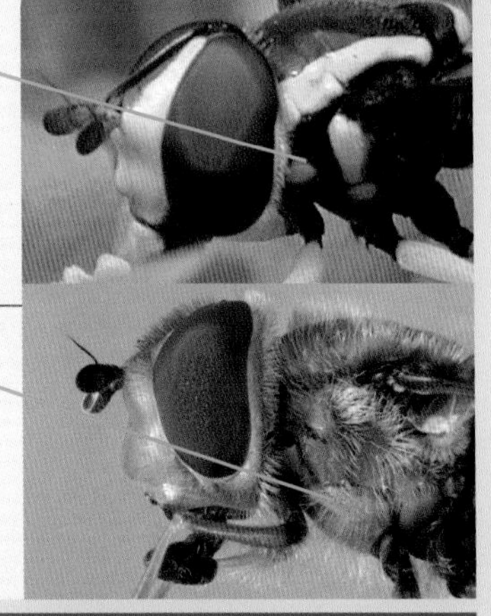

b) Sides of the thorax strongly hairy or dusted, with no distinct yellow markings

The sides of the thorax may have patches of murky yellowish or olive dusting, but not sharply-defined citrus-yellow markings.

2

from **1a**

a) Abdomen broad:

Long, porrect antennae
Chrysotoxum *pp. 98–103* – pictured right

'Normal' antennae
Xanthogramma *p. 104* – pictured 1a above

b) Abdomen narrow:

Large with darkened wings
Doros *p. 106*

Small with clear wings
Sphaerophoria *pp. 108–111* – pictured right

3 from **1b**

a) Bumblebee mimic:

Scutellum yellow

Eriozona syrphoides p. 112

b) Abdomen with large creamy white or bluish rectangular markings on the 2nd segment, much bigger than any markings of the rest of the abdomen

Leucozona pp. 112–115

c) Unconvincing wasp mimics; black abdomens with yellow, orange or creamy-yellow markings:

Wing with obvious long, black stigma

Dasysyrphus pp. 116–119

Wing with less obvious usually brown or yellow stigma ▶ 4

4 from **3c**

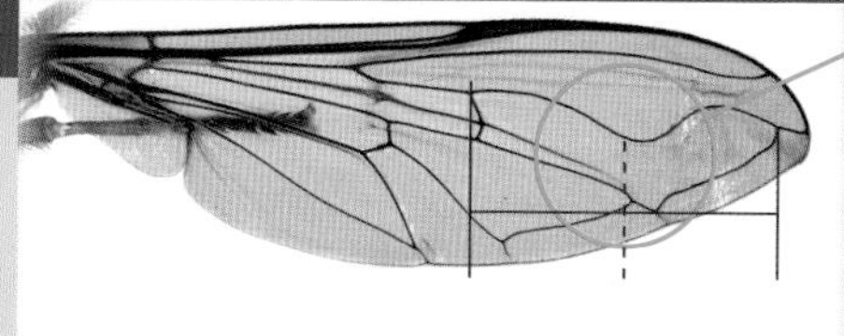

a) Wing with vein R_{4+5} distinctly dipped; basal half of discal cell parallel-sided or even narrowing

Note: the sub-genus *Lapposyrphus* (within *Eupeodes*) also exhibits this character, but is not covered by this book.

Didea p. 120
Megasyrphus erraticus p. 144

b) Wing with vein R_{4+5} not dipped; discal cell steadily widens towards wing edge ▶ 5

5 from **4b**

To establish the identification of the 'black-and-yellow' genera that do not have a dip in vein R_{4+5} is not completely straightforward. Markings on the abdomen are helpful in some cases, but some patterns occur in several genera. These genera can only be separated reliably by features that are hard to see without strong magnification.

The table overleaf details the most apparent characteristics, as well as an indication of any reliable features for identification:

Episyrphus p. 138
Eupeodes p. 124
Syrphus p. 150
Epistrophe p. 140
Parasyrphus p. 146
Melangyna p. 128
Meligramma p. 134
Scaeva p. 122
Meliscaeva p. 136

6 Simplified key to the 'black-and-yellow' Syrphini genera that have wing vein R_{4+5} not dipped and lack defined yellow markings on the side of the thorax

Genus	Abdomen markings	Face	Additional identification pointers
Episyrphus *p. 138*	Orange bands are completely or partially split into two bands by narrow black bars.	Yellow–dark, variable	**Abdomen pattern** diagnostic The 'marmalade fly', whose divided orange bands are supposed to look like thick-cut marmalade!
Eupeodes *pp. 124–127*	Yellow T3+4: some banded; some lunulate	Yellow Beware – some *Eupeodes* may have faint black marks on the lower part of the face	**Edges of abdomen with completely black hairs on T3+T4** diagnostic
Syrphus *pp. 150–153*	Yellow T3+4: banded	Yellow	**Hairs on upper surface of squamae diagnostic**
Epistrophe *pp. 140–145*	Yellow T3+4: banded	Yellow	Beware – some spp. very similar to *Syrphus* and *Parasyrphus* – see species accounts for details.
Parasyrphus *pp. 146–149*	Yellow T3+4: moustache-shaped bands except *P. punctulatus* ♂ – 'golf-club' spots ♀ – paired spots	With partial or complete black stripe	Face colour and lack of hairs on squamae (cf. *Syrphus*) Beware – some spp. very similar to *Syrphus*.
Melangyna *pp. 128–133*	Yellow–cream T3+4: oval–squarish Beware – ♀ *M. quadrimaculata* has no markings	With partial or complete black stripe	**Face 'knob' darkened**; relatively small and narrow hoverflies with mainly black legs
Meligramma *p. 134*	Yellow–cream T2: triangular T3+4: oval–squarish	With partial or complete black stripe	**Face 'knob' yellow** **Facial features + abdomen markings** diagnostic
Scaeva *p. 122*	White or pale yellow T3+4: lunulate	With partial or complete black stripe	**Swollen frons** diagnostic Beware – some species of *Dasysyrphus* have narrow comma-shaped markings, but they also have an extended stigma (see **3c**).
Meliscaeva *p. 136*	Yellow T3+4: *M. cinctella* – banded *M. auricollis* – squarish spots but variable	With partial or complete black stripe	Wings somewhat elongate

Images illustrating the key characteristics of the 'black-and-yellow' Syrphini genera

Face: with partial or complete black stripe (*Parasyrphus*)

Face: yellow (*Epistrophe*)

Abdomen markings:

LEFT: The unique orange-banded pattern of *Episyrphus balteatus*

RIGHT: *Eupeodes* abdomen showing segments T2–T4

BELOW LEFT: **Banded** – typically with T3+4 banded and T2 separated

BELOW CENTRE: **Oval-squarish** markings on T3+4

BELOW RIGHT: **Lunulate** or comma-shaped markings on T3+4

Syrphus sp.

Melangyna umbellatarum

Scaeva pyrastri

Chrysotoxum

8 British species (6 illustrated)

Large and very smart wasp mimics with long antennae. Very little is known about larval biology, although they are believed to feed on root aphids in ants' nests. The genus splits into three groups: *C. bicinctum*, which has very distinctive markings; *C. festivum* and *C. vernale*, where the markings on the abdomen do not go all the way to the edges, leaving a black 'bead' down the sides; and the rest (*C. arcuatum*, *C. cautum*, *C. elegans*, *C. octomaculatum* and *C. verralli*), which can be quite difficult to separate and are therefore referred to as the 'difficult five' – see *page 100*.

Chrysotoxum bicinctum

Wing length: 7–10·25 mm

Identification: The distinctive yellow bars on abdomen segments T2 and T4, combined with the chocolate-coloured wing markings are unmistakable.

Similar species: It should not therefore be possible to confuse this species with any other British hoverfly, however *Dasysyrphus tricinctus* (*p. 116*) might cause confusion at some angles in photographs, although the strong shading along the wing margin in *C. bicinctum* should be distinctive.

Observation tips: Found in open grasslands and grassy woodland rides. Common in southern England; less common in lowland areas of the north; occasionally found as far north as the north coast of Scotland.

'BEADED' *CHRYSOTOXUM*

Chrysotoxum festivum

Wing length: 8·25–12 mm

Identification: The distinct downturned bars on abdomen segments T2–T4, together with its size and wholly orange legs (or partially black on the front femora only) make this species quite distinctive.

Similar species: It can only be confused with the Endangered and very rare *C. vernale* (not illustrated), which is confined to very restricted parts of southern Hampshire and Dorset. The differences are subtle and involve the shape of the abdomen markings and the extent of dark markings on the legs – features which are both rather variable.

Observation tips: Predominantly a southern species, but occurs into southern Lancashire with a scattering of records north to Scotland. There are indications that this species is spreading northwards in response to climate change, and it should therefore be watched for in lowland northern localities.

Chrysotoxum bicinctum ♀ ×**6**

Chrysotoxum festivum ♂ ×**6**

Similar species: The 'difficult five' are: *Chrysotoxum arcuatum* and *C. cautum* (*below*); *C. elegans* and *C. verralli* (*p. 102*); and the Endangered *C. octomaculatum* (not illustrated) which is confined to the heaths of Dorset and Surrey. See table on *page 102* for a summary of the differences between these species.

Chrysotoxum arcuatum

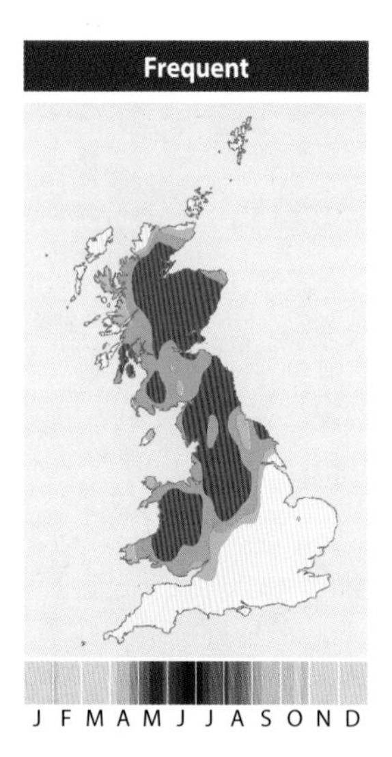

Wing length: 8–10·25 mm

Identification: This is the smallest member of the 'difficult five' but is relatively straightforward to identify because the length of the antennal segment 3 is considerably longer than segments 1 and 2 added together. In the field, its abdomen appears to be more globular in shape than in the other species.

Observation tips: A northern and western species which is most frequently found in coniferous woodlands north of a line between the Humber and the Severn. It is also one of a suite of northern species that occur in the Norfolk Brecklands.

Chrysotoxum cautum

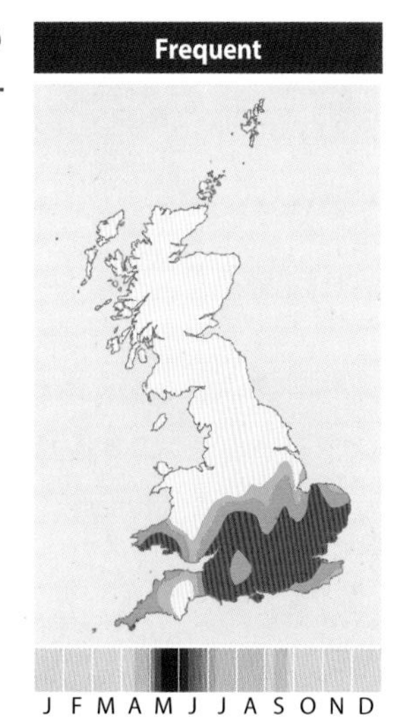

Wing length: 10·25–13 mm

Identification: This is one of the biggest and bulkiest of the 'difficult five'. Both sexes can be identified on the basis of the ratio of the length of the 1st and 2nd antennal segments: segment 1 is about ⅔ the length of segment 2. Males are easier to distinguish from other *Chrysotoxum* as they have an extremely large and obvious genital capsule.

Observation tips: A predominantly southern species which occurs as far west as Pembrokeshire in South Wales and as far north as Lincolnshire in the east. During peak emergence in late May and early June, it can be found in grasslands and open sunny rides.

Chrysotoxum species not otherwise covered:

C. octomaculatum [EN] – UK Biodiversity Action Plan (UKBAP) priority species. Extremely rare. The few records are from the fringes of heathlands in Surrey and Dorset.

C. vernale [EN] – Very rare species recorded from the coast of Hampshire and Dorset, especially the latter.

C. arcuatum
length seg. 1 + seg. 2 < length seg. 3

Chrysotoxum arcuatum ♀ ×**6**

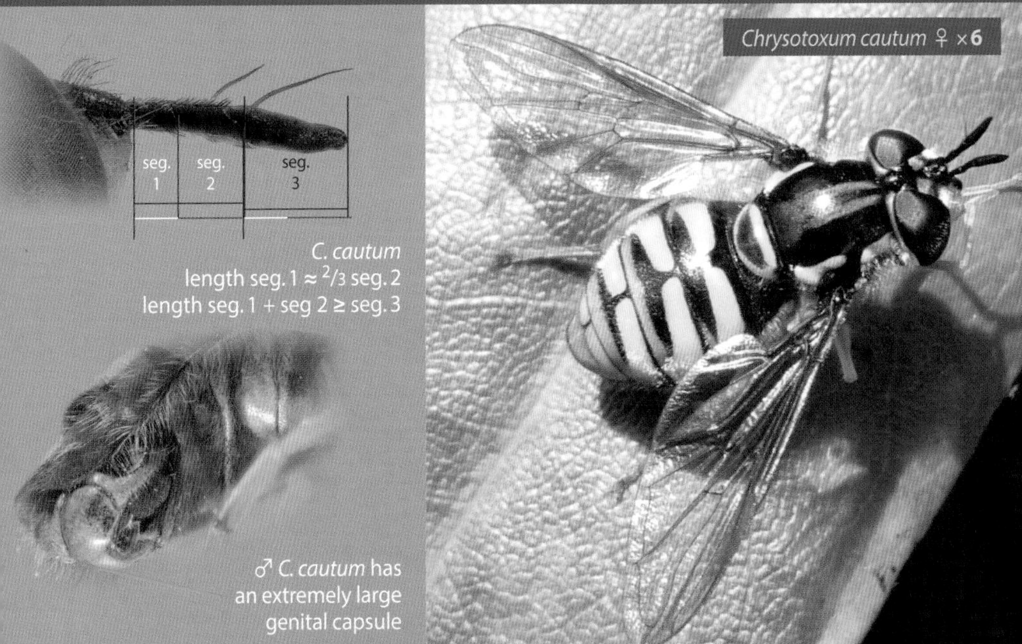

Chrysotoxum cautum ♀ ×**6**

C. cautum
length seg. 1 ≈ $^2/_3$ seg. 2
length seg. 1 + seg 2 ≥ seg. 3

♂ *C. cautum* has an extremely large genital capsule

Chrysotoxum elegans

Wing length: 9·5–12 mm

Identification: One of the two largest of the 'difficult five' (with *C. cautum*), this is one of the most tricky to identify and demands careful examination under magnification. The main character used to separate it from *C. verralli* and *C. octomaculatum* is the shape of the yellow bars on the abdomen, which are well separated from the front edges of the segments by a black band which widens towards the sides.

Observation tips: Although it does occur inland, predominantly on chalk downland in eastern England, the majority of records are from the coast of south-west England and South Wales. It has a longer flight period than *C. cautum* and is, therefore, the more likely of these species to be encountered into August.

Chrysotoxum verralli

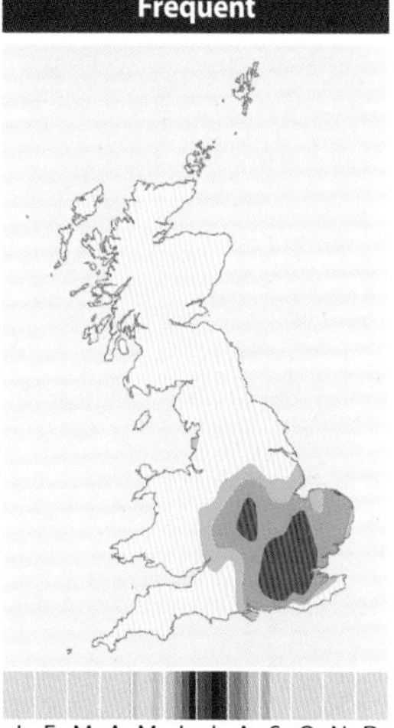

Wing length: 8·25–10·5 mm

Identification: Generally much smaller than *C. cautum* and *C. elegans*, but tricky to identify. Difficult to separate from the extremely rare *C. octomaculatum* (from which it was only split in the mid-20th Century). The yellow markings on the abdomen are more extensive than those of *C. elegans* so that the black front edges are narrower and more parallel-sided, broadening only at the outer corners.

Observation tips: Occurs mainly in south-east England but also on the north-west coast from Liverpool to Morecambe Bay. A mid-Summer species which is most frequently found in hot, open grasslands or heathy rides where it will visit flowers such as Wild Parsnip and Marjoram.

Separating the 'difficult five' British *Chrysotoxum* species requires careful examination.

SPECIES	FEATURES	
C. cautum	ANTENNA: **seg. 1 about ⅔ seg. 2; length seg. 1 + seg. 2 ≥ seg. 3**	
C. arcuatum	ANTENNA: seg. 1 = seg. 2; **length seg. 1 + seg. 2 < seg. 3**	
C. elegans	ANTENNA: seg. 1 = seg. 2; **seg. 1 + seg. 2 > seg. 3**	ABDOMEN: black bands **widen towards sides**
C. verralli	ANTENNA: seg. 1 = seg. 2; **seg. 1 + seg. 2 ≈ seg. 3**	Separation of these species is difficult. Subtle differences in the abdomen patterns require comparison with known specimens. Records of *C. octomaculatum* require independent verification.
C. octomaculatum [EN]		

Chrysotoxum elegans ♀ ×**6**

Chrysotoxum verralli ♀ ×**6**

Xanthogramma

3 British species (2 illustrated)

Very smart black-and-yellow hoverflies which have yellow faces, a yellow stripe running along the edge of the top surface of the thorax, and yellow markings on the sides of the thorax. They could only be confused with *Chrysotoxum* (*p. 98–103*) but lack the long, forward-projecting antennae of that genus. The larvae are believed to live in the nests of yellow or black ants of the genus *Lasius*.

Xanthogramma citrofasciatum

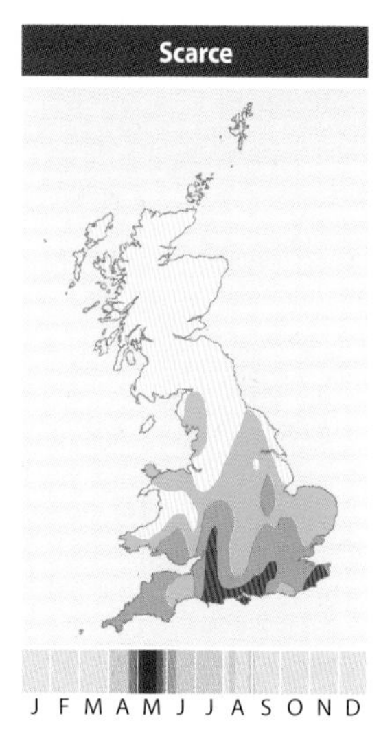

Wing length: 6·5–10·25 mm

Identification: The sharply defined, narrow, triangular marking on abdomen segment T2, together with the lemon-yellow markings makes this species quite easy to recognise.

Similar species: Other *Xanthogramma* species. The legs of *X. citrofasciatum* are predominantly orange, rather than yellow, and lack the black markings that are present on the hind legs of the other two species.

Observation tips: Short grassland with abundant Yellow Meadow Ant (*Lasius flavus*) nests in warm situations (south-facing slopes, sheltered glades in scrub, *etc.*). This includes acid grasslands, some coastal grazing marshes and chalk grasslands. A scarce species with a predominantly southern distribution, reaching as far north as the Lake District.

Xanthogramma pedissequum

Wing length: 7·25–9·75 mm

Identification: Much more orange-yellow than *X. citrofasciatum* with much broader marking on abdomen segment T2. The extent of yellow markings on the sides of the thorax has been used to segregate a complex of species. So far, we know of at least one additional British species, *X. stackelbergi*. Photographs of the thoracic markings and/or a specimen should therefore be kept to ensure confirmation of records.

Similar species: Other *Xanthogramma* species.

Observation tips: Predominantly a grassland species, but it also occurs in woodland rides (although these may refer to *X. stackelbergi* which appears to be a woodland edge species). Mainly southern, where it is widespread, but there are a few scattered records into central Scotland.

Xanthogramma species not otherwise covered:

X. stackelbergi – A recent addition to the British list (2012). The very few records so far suggest that it is less frequent than *X. pedissequum* and perhaps associated with less open habitats – *e.g.* woodland rides and edges.

Xanthogramma citrofasciatum ♂ ×**7**

Xanthogramma pedissequum ♀ ×**7**

Doros

1 British species (illustrated)

A large, narrow-waisted hoverfly with a large, chocolate-coloured marking along the front edge of the wing. It should not be mistaken for any other species of hoverfly, but could be confused initially with members of the thick-headed fly genus *Conops* (Conopidae). This is a highly enigmatic species which has been the subject of concerted effort as part of the UK Biodiversity Action Plan (UKBAP). Despite this effort, very little is known of its larval biology – though the larvae are suspected to be associated with aphids tended by the black ant *Lasius fuliginosus*.

Doros profuges

BAP: Near Threatened

Scarce

Wing length: 11·25–13·25 mm

Identification: Unmistakable, due to its size and wing markings (described above).

Similar species: This species cannot be confused with other British hoverflies, but might be confused with the genus *Ceriana* elsewhere in Europe. Confusion with the genus *Conops* (family Conopidae) (shown below) is also likely.

Observation tips: Extraordinarily difficult to find, even at its few well-known sites. This may be, in part, because the flight period is very short, making it easy to miss. It appears to be most frequently noted from scrubby chalk downland and limestone grassland. Although most records come from southern England, the most consistent site in recent times has been a locality in Silverdale, Lancashire. It was also caught in a Malaise trap on the island of Mull in 1991.

A conopid fly, with which *Doros* could potentially be confused.

Doros profuges ♀ ×**7**

Sphaerophoria

11 British species (3 illustrated)

Small, elongate, black-and-yellow banded hoverflies with yellow faces, a yellow scutellum and prominent yellow markings on the side of the thorax. Males have a large genital capsule which forms a bulge under the end of the abdomen. **Definite identification is only possible in males based on genital characteristics. Females cannot be identified** to species but are often mistaken for (and reported as) other narrow-banded hoverflies such as smaller *Parasyrphus* (*pp. 146–149*). The larvae feed on a variety of ground-layer aphids.

Sphaerophoria interrupta

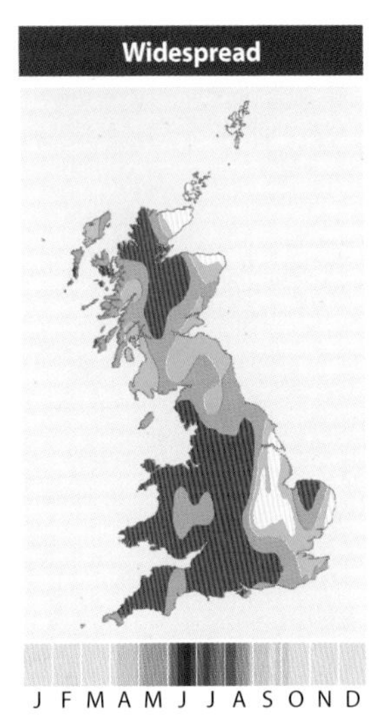

Wing length: 4·75–6·5 mm

Identification: Only identifiable using microscopic characters of the male genitalia, this species having a thumb-like process. It is included here to represent the six species with paired spots on the abdomen segments. Other species have banded abdomens.

Similar species: Most other *Sphaerophoria,* including *S. scripta* (*p. 110*) which have developed in cold conditions and are therefore darker in colour and may have spots rather than bands on the abdomen segments.

Observation tips: A common and widespread grassland species which occurs in a wide range of situations, often in damper locations. It can be found with several other *Sphaerophoria* species, most notably *S. fatarum* and *S. philanthus.*

Sphaerophoria species not otherwise covered:

ABDOMEN: **markings paired spots**

S. bankowskae [DD]	– Only 3 known records from Essex, Northamptonshire and the Scottish Highlands.
S. fatarum	– One of the commonest species of the genus on open heathland and moorland in Scotland with a scattering of records farther south.
S. philanthus	– Widespread in the north and west on dry heaths and moorland, also in heathy rides in conifer woods.
S. potentillae [VU]	– Discovered in Devon in 1989 and known only from a few Culm grassland sites.
S. virgata [NS]	– A scarce species of heathland and moorland with most records from central Scotland and the heaths of southern England.

ABDOMEN: **distinctive banded pattern; slightly waisted**

S. loewi [NT]	– Most similar to *S. rueppellii*, this is a rare, mainly coastal species with the majority of records from brackish reedbeds.

ABDOMEN: **usually banded**

S. batava	– Widely scattered records from heathland and open rides in woodland.
S. taeniata	– Mainly found in southern England, generally in rich, unimproved grassland.

Sphaerophoria interrupta ♂ ×**10**

Sphaerophoria ♀ ×**8**

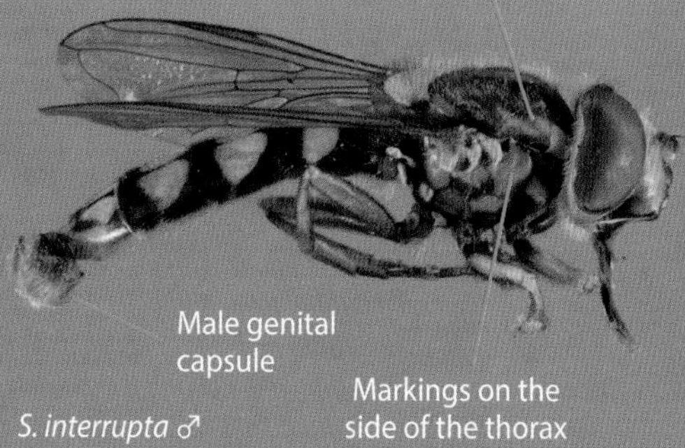

S. interrupta ♂

Sphaerophoria male genitalia

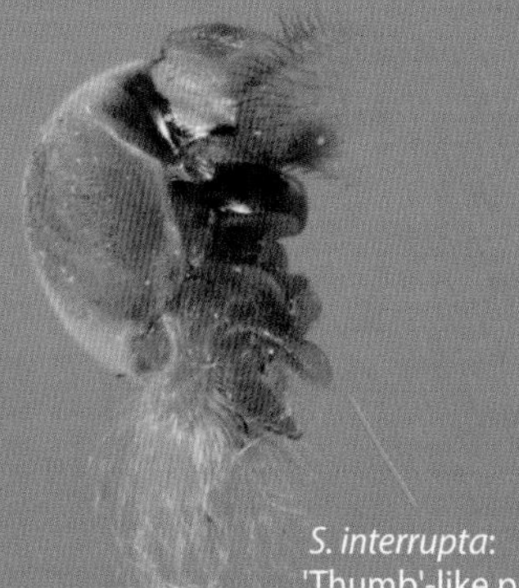

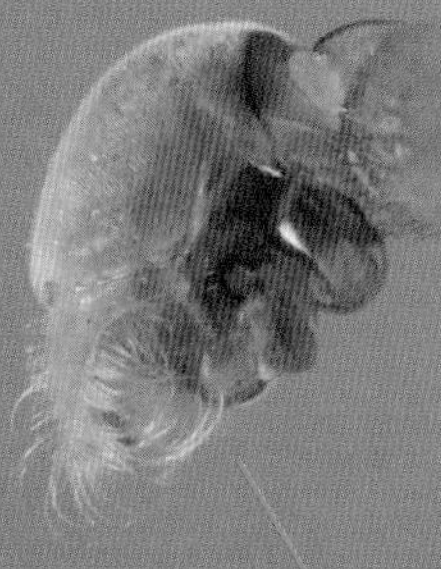

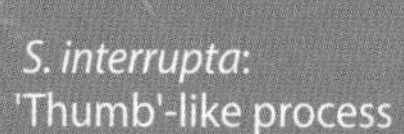

S. interrupta: 'Thumb'-like process

S. rueppellii: Very open and hairy

S. scripta: characteristic inner process shape

Sphaerophoria rueppellii

Frequent

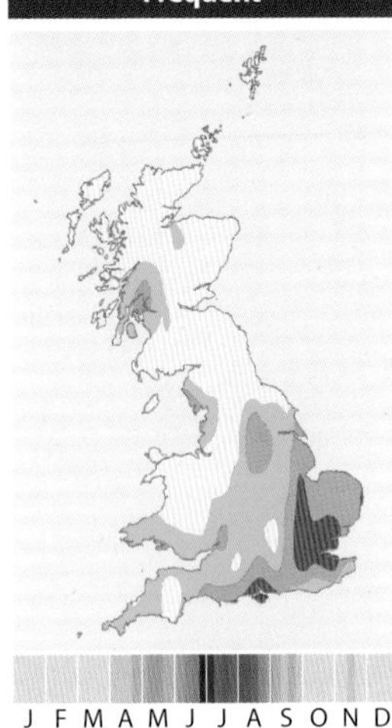

Wing length: 4·25–6·5 mm

Identification: Males of this species are reasonably straightforward to identify if examined carefully. The combination of yellow antennae, an interrupted thoracic stripe and a slightly waisted abdomen is fairly distinctive. The Near Threatened *S. loewi* (not illustrated) is similar, but has black antennae. Genitalia must be checked for definite identification.

Similar species: *Sphaerophoria loewi* (see above).

Observation tips: A southern and coastal species which is particularly abundant around the Thames Estuary. Inland records are less frequent, but it can be found in situations like field corners where plants such as Scentless Mayweed, Fat-hen and knotweeds are abundant.

See *page 109* for details of the male genitalia.

Sphaerophoria scripta 

Widespread

J F M A M J J A S O N D

Wing length: 5–7 mm

Identification: Reasonably distinctive because the abdomen is longer than the wings, and protrudes beyond the wing-tips when the hoverfly is at rest. The markings on the abdomen are usually broad yellow bands. However, the colouration is strongly influenced by the temperature at which larvae developed. Early Spring individuals are often darker and may have spotted rather than banded abdomens. There are two other banded species (*S. batava* and *S. taeniata*), which require microscopic examination of the male genitalia for identification.

Similar species: *Sphaerophoria batava* and *S. taeniata* in particular (see above), but *S. scripta* can have spots when it has developed in colder environments (especially in the Spring), so confusion with *S. interrupta*-type species is possible. Longer black-and-yellow male *Platycheirus* (*pp. 78–91*), male *Melanostoma* (*p. 76*) and possibly *Melangyna cincta* (*p. 132*) and *Meliscaeva cinctella* (*p. 136*).

Observation tips: A partial migrant, which probably does not have a resident population in the more northerly parts of Britain. It is the commonest *Sphaerophoria* and is widespread in grasslands in England and Wales.

See *page 109* for details of the male genitalia.

Sphaerophoria rueppellii ♂ × **10**

Sphaerophoria scripta ♂ × **10**

S. scripta ♂ (×5) with spots rather than connected band type markings

Eriozona

1 British species (illustrated)

A bumblebee mimic. The larvae feed on tree-dwelling aphids

Eriozona syrphoides

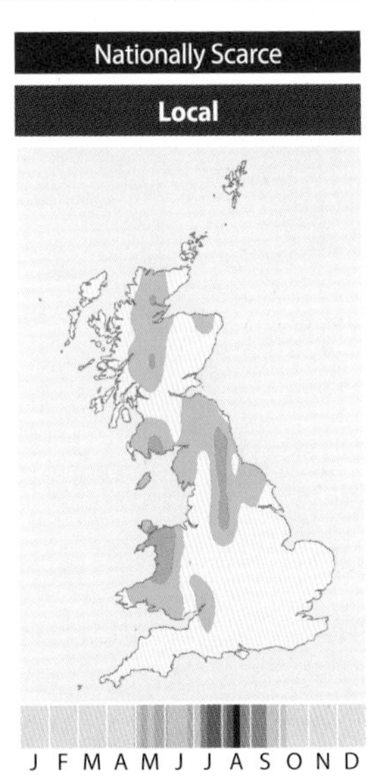

Wing length: 12 mm

Identification: This is a distinctive bumblebee mimic due to its wholly yellow face (unique among British bumblebee mimics) and a yellow scutellum (black in *Cheilosia illustrata*).

Similar species: *Cheilosia illustrata* (*p. 162*) is frequently mistaken for this species. There may also be confusion with *Eristalis intricaria* (*p. 206*) and *Leucozona lucorum*.

Observation tips: A northern and western species which occurs in conifer plantations. In Europe, the larvae are reported to feed on aphids on Spruce. It is usually found in mid- to late Summer in open, sunny rides where it visits a variety of flowers such as Hogweed, Angelica, scabiouses and Heather. It is a relatively recent colonist that was first found in Wales in the 1960s and has subsequently increased its range, although is nowhere common.

See *Identifying wasp and bee mimics* on *pages 64–66*.

Leucozona

3 British species (all illustrated)

Species in this genus differ in colour and markings from the majority of other hoverflies, the blue-grey markings of female *L. glaucia* being particularly unusual. They are all woodland species and can be found together on Hogweed flowers. The larvae mainly feed on ground-layer aphids, although *L. lucorum* larvae have also been found feeding on arboreal aphids.

Leucozona lucorum

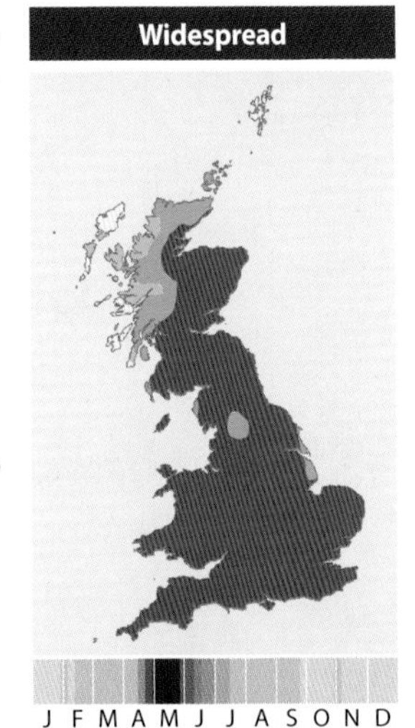

Wing length: 7·75–10 mm

Identification: Predominantly a Spring hoverfly with broad creamy markings on the abdomen segment T2 and distinct wing clouds. Males tend to be darker than females. Most likely to be confused with *Cheilosia illustrata*, but in that species the the scutellum is black rather than yellow and the pale areas are due to white hair, not markings.

Similar species: *Eriozona syrphoides*, *Cheilosia illustrata* (*p. 162*), *Eristalis intricaria* (*p. 206*) and *Volucella pellucens* (*p. 246*).

Observation tips: Widespread in woodland rides and along woodland edges and hedgerows. Visits low-growing flowers such as Greater Stitchwort and Garlic Mustard in dappled sunshine. Although mainly a Spring species, it sometimes has a much less abundant second generation in mid-Summer.

Eriozona: SYRPHINI

Eriozona syrphoides ♀ ×**6**

Leucozona: SYRPHINI

♂ – *Leucozona lucorum* ×**5** – ♀

Leucozona glaucia

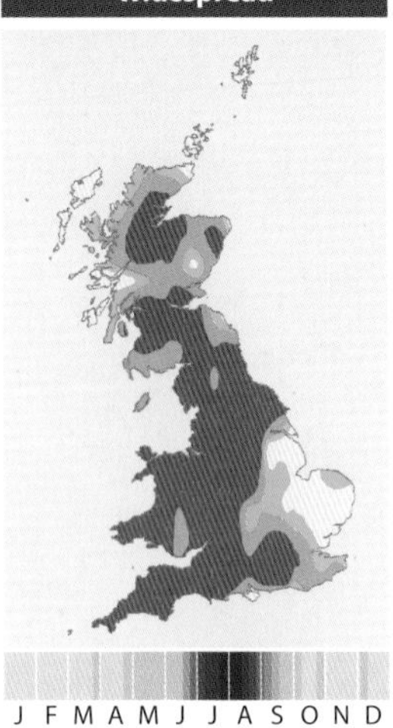

Wing length: 8–11·25 mm

Identification: The blue-grey abdominal markings of females are unmistakable. Males can be separated from *L. laternaria* by the yellow scutellum and their front legs are also mainly yellow (extensively darkened in *L. laternaria*).

Similar species: This species can really only be confused with *Leucozona laternaria*.

Observation tips: Mainly a western and northern species, but also widespread in the wooded counties of south-east England. It is, however, scarce or absent in East Anglia and the East Midlands. Particularly abundant in the north, where it can be one of the commonest hoverflies in mid-Summer on Hogweed and Angelica in woodland rides.

Leucozona laternaria

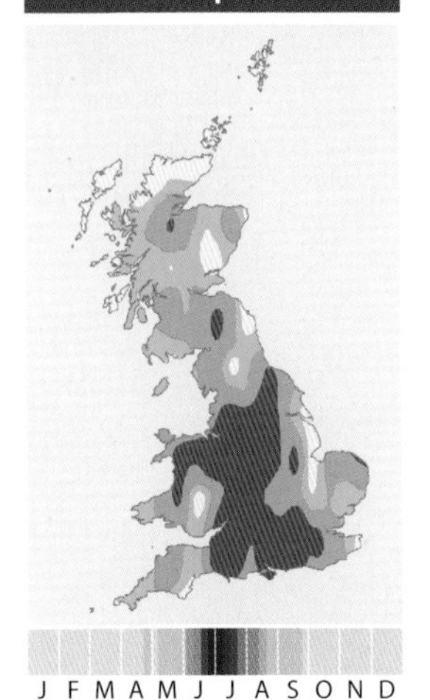

Wing length: 7–10 mm 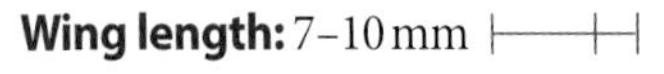

Identification: Very similar to male *L. glaucia* but can be readily separated by the darkened scutellum and front legs (yellow in *L. glaucia*).

Similar species: This species can really only be confused with *Leucozona glaucia*.

Observation tips: A woodland species which occurs most frequently on umbellifers, especially Hogweed and Angelica. Its larvae are particularly associated with aphids on umbellifers. It is slightly more widespread in eastern England than *L. glaucia*, but is still less abundant than it is further west.

♂ – *Leucozona glaucia* × **5** – ♀

Leucozona species can be separated as follows:

***L. lucorum*:**
ABDOMEN: Distinctive, with broad, creamy markings on T2
WING: distinct clouds

***L. glaucia*:**
♂ similar to *L. laternaria*
♀ ABDOMEN: Unmistakable blue-grey markings
SCUTELLUM: **yellow**
FRONT LEGS: **yellow**

***L. laternaria*:**
similar to *L. glaucia* but
SCUTELLUM: **darkened**
FRONT LEGS: **darkened**

Leucozona laternaria ♀ × **5**

Dasysyrphus

7 British species (4 illustrated)

Yellow-and-black hoverflies with hairy eyes that normally have a central black stripe on the face and a **distinct, elongate black stigma** (the marking along the front edge of the wing). Several of the species have distinctive markings that make them straightforward to identify. However, there is a group of species which includes *D. venustus* (see *page 118*) that is much more challenging. This group is being revised and, once published, there will be several additions to the British list. For this reason, it is recommended that specimens are retained. The larvae are camouflaged with bark-like colour patterns. They spend the day resting on twigs and branches and the night feeding on aphids on both coniferous and deciduous trees.

Dasysyrphus albostriatus

Wing length: 6·25–9 mm

Identification: The downward-facing oblique bars on abdomen segments T2–T4 and the pair of grey stripes on the thorax make this species straightforward to identify.

Similar species: *Didea* (*p. 120*), arguably, have similar downward-pointing markings on abdomen segments T2 and T3, but the markings are broader; the overall form of *Didea* is broader and more robust, and the wing venation is different.

Observation tips: There are two distinct generations of this widespread woodland species, one in the Spring and another in late Summer and early Autumn. Both sexes visit flowers and bask on sunlit leaves along woodland rides.

Dasysyrphus tricinctus

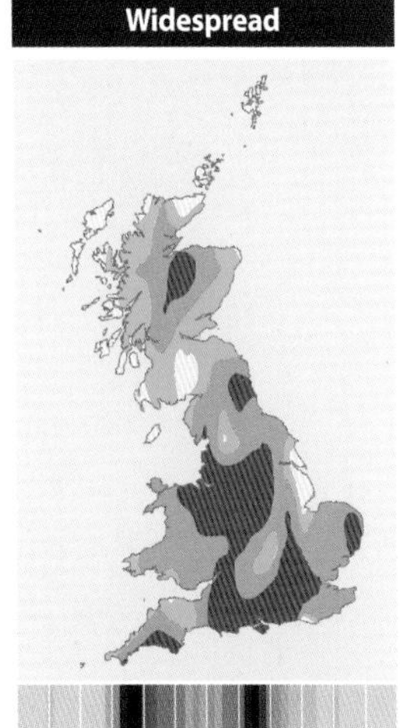

Wing length: 7·25–10·25 mm

Identification: The abdominal markings consist of strong bars on segment T3 and narrow bars on segment T4. This pattern is relatively distinct and should make it readily identifiable.

Similar species: Confusion with *Chrysotoxum bicinctum* (*p. 98*) might occur under some circumstances but that species has extensive darkening along the leading edge of the wing and yellow markings on the thorax side which should eliminate any confusion.

Observation tips: A woodland species which is most frequently encountered in coniferous woodlands and on heathlands. Adults are flower visitors, favouring open yellow flowers such as buttercups and Dandelion.

Dasysyrphus albostriatus ♂ ×**6**

Dasysyrphus tricinctus ♂ ×**6**

Dasysyrphus venustus

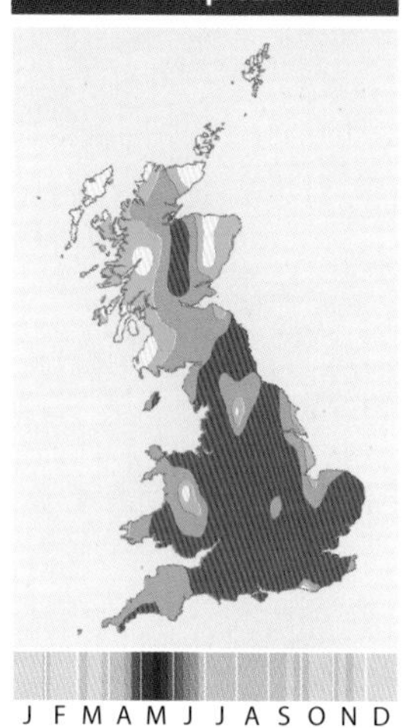

Wing length: 6·25–10 mm

Identification: The narrow, hooked bars on abdomen segments T3 and T4 are characteristic of a number of very similar species. In *D. venustus* the bars reach the edge of the abdomen whereas those of *D. pinastri* and *D. pauxillus* are separated by a narrow black strip along the edge. *D. hilaris* is very similar but has a completely yellow face. *D. friuliensis* is also similar but has much more strongly hooked bars and completely dark antennae (otherwise it is somewhat lighter than *D. venustus*, especially underneath). Bearing in mind its close similarity to these other species, it cannot be reliably identified from photographs. Unpublished studies show that '*D. venustus*' is a species complex which includes several more species, at least two of which occur in Britain.

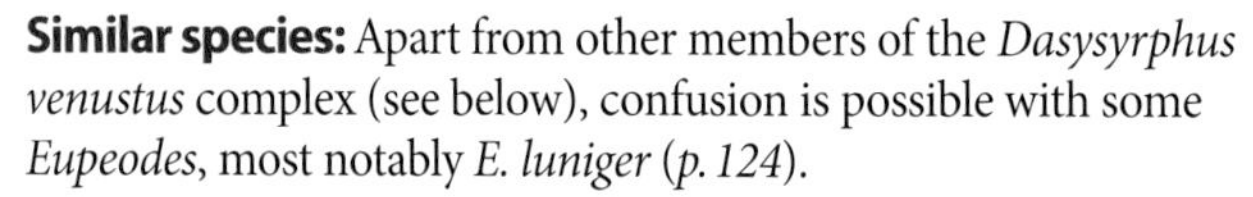

Similar species: Apart from other members of the *Dasysyrphus venustus* complex (see below), confusion is possible with some *Eupeodes*, most notably *E. luniger* (*p. 124*).

Observation tips: A woodland species which occurs in both deciduous and coniferous woodlands across lowland Britain but is much commoner in the south. It basks on sunny leaves and is a regular visitor to flowers such as buttercups and *Acer* species, especially Field Maple.

Dasysyrphus species not otherwise covered:

D. friuliensis/ *D. hilaris*	– These are part of a complex of species around *D. venustus* which is still under review – so there may well be more species split off.
D. pinastri/ *D. pauxillus*	– The *pinastri* complex is a widespread, but apparently declining assemblage associated with conifer plantations. Although it occurs throughout GB, it is more frequent in northern England and Scotland.

D. friuliensis (1979) and *D. pauxillus* (2010) are both recent additions to the British list and are tricky to separate from *D. venustus* and *D. pinastri* respectively. There are very few records of either species.

Dasysyrphus pinastri ♀ ×6

The yellow bars on the abdomen of *D. pinastri* do not reach the edge of the abdomen.

▲ ♂ – *Dasysyrphus venustus* ×6 – ♀ ▼

Didea

3 British species (2 illustrated)

This is a highly distinctive genus, for two reasons. Firstly, **the loop in wing vein R_{4+5} is quite distinct** (although not as strongly developed as in the Eristalini (*p. 198*)); this is an unusual feature amongst the yellow-and-black hoverflies. Secondly, the oblique abdominal markings give these flies a very distinct and 'aggressive' appearance. The larvae feed mainly on aphids on conifers. Only one species, *Didea fasciata*, is at all abundant.

Didea fasciata

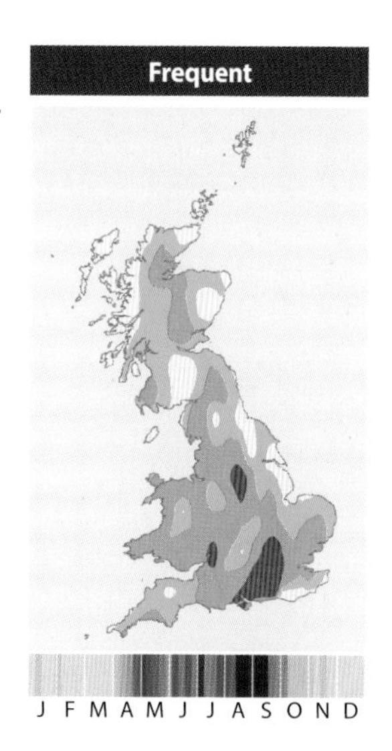

Wing length: 8·25–11 mm

Identification: Black and yellow hoverfly with distinctive abdominal markings. The only species with which *D. fasciata* could be confused is *D. intermedia*. Separation should be straightforward on the basis of *D. fasciata*'s yellow haltere club (black in *D. intermedia*) and the greater extent of yellow on the scutellum of *D. fasciata*.

Observation tips: A woodland hoverfly which has two marked peaks of emergence: one in the Spring and another in late Summer. It occurs throughout mainland Britain but is predominantly a southern species. Males are strongly territorial and can be seen defending sunlit leaves.

Didea intermedia

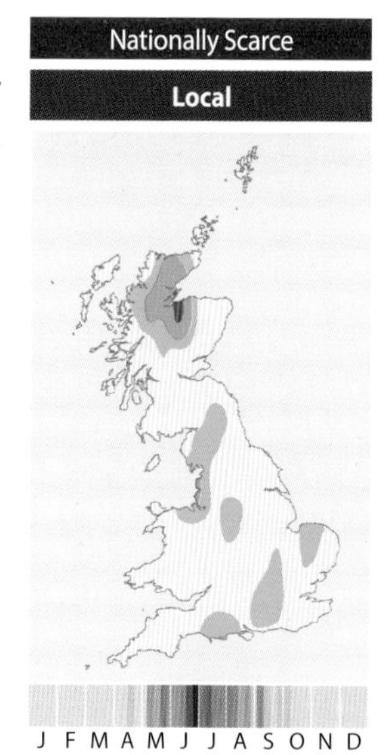

Wing length: 7–10 mm

Identification: Very similar to *D. fasciata* but can be distinguished by the black haltere club – yellow in *D. fasciata*. The scutellum of *D. intermedia* is also much more extensively darkened along the rim than *D. fasciata*.

Observation tips: Adults mainly fly from June to August and seem not to stray far from conifers. They can be found on yellow composites in coniferised heathland in southern England, and in conifer plantations in the north and west. In Scotland, it can be quite abundant along rides in coniferous woodlands and plantations where it frequently visits the white flowers of Heath Bedstraw.

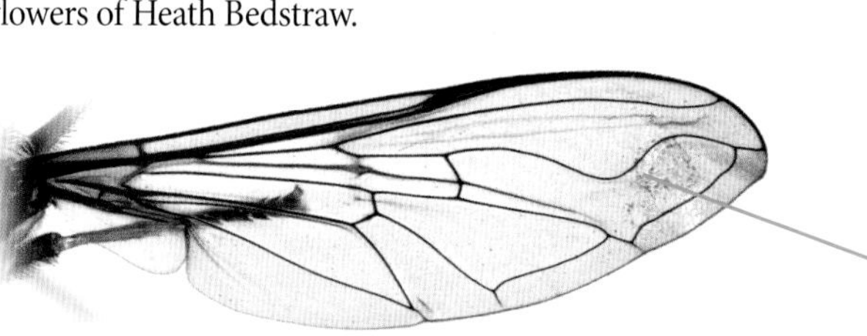

Didea wing showing the distinct loop in vein R_{4+5}.

Didea species not otherwise covered:

D. alneti – A very rare vagrant with abdomen markings that are usually greenish or turquoise, rather than yellow. The last known records were from Northumberland in 1989.

Didea fasciata ♀ ×**6**

Didea intermedia ♀ ×**5**

D. fasciata – yellow

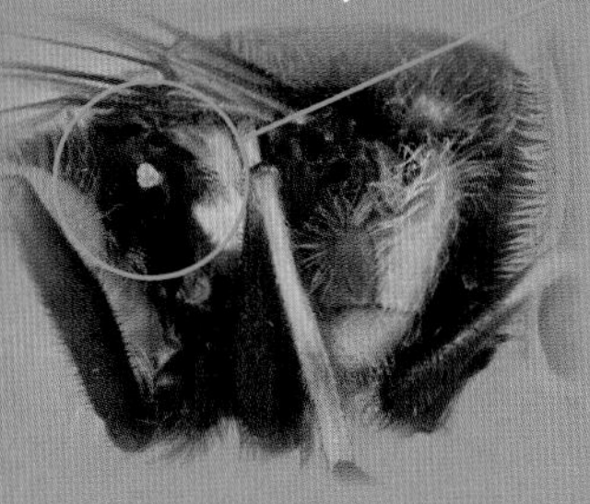

The colour of the haltere club is diagnostic.
This character is often hidden under the wings and will not necessarily be shown by a photograph.

D. intermedia – black

Similar species:
The two resident *Didea* species are similar, but can be readily separated. *Dasysyrphus albostriatus* (*p. 116*) has similar downward-pointed abdominal markings but these are narrower and the insect has a more delicate build.

Scaeva

5 British species (2 illustrated)

Black and white or yellow marked hoverflies with hairy eyes, and in which the area between the eyes bulges strongly (more so than in any other British hoverfly). Whilst *S. selenitica* may be resident, the commonest species, *S. pyrastri*, is a migrant that is not permanently resident in Britain. The other three species (*S. albomaculata*, *S. mecogramma* and *S. dignota*) are rare vagrants with only a handful of records. The larvae feed on aphids.

Scaeva selenitica

Wing length: 10·5–12 mm

Identification: Similar to *S. pyrastri*, but the abdominal markings are pale yellow. However, it is quite variable, making positive identification tricky on occasion. It is worth bearing in mind that *S. dignota*, an extremely similar European species, has occurred as a vagrant to Britain.

Similar species: *Scaeva dignota*, which was confirmed to occur in Britain in 2013, is very similar: the characters separating these two species are extremely subtle and call for comparison with voucher material. In addition, confusion is possible with *Eupeodes lundbecki* (rare vagrant – not illustrated), *S. albomaculata* (rare vagrant – not illustrated) and *S. pyrastri*.

Observation tips: Most frequent in the vicinity of coniferous woodland or on coniferised heathland, suggesting that there is a resident population – although it may be augmented by migrants. The larvae have been found feeding on aphids on pines. Widespread, but scarce and most often found on heathlands in southern England.

Scaeva pyrastri

Wing length: 9·25–12·5 mm 

Identification: A relatively large hoverfly with white, comma-shaped spots (yellow in *S. selenitica*) on a black background. This, in combination with the inflated frons, makes it quite unmistakable. Completely black individuals can occur, but are usually recognisable by the inflated frons.

Similar species: *Scaeva albomaculata* (a rare vagrant) and *S. selenitica*.

Observation tips: A migratory species which, like Red Admiral or Clouded Yellow butterflies, arrives in Britain in highly variable numbers; in some years it is almost absent, but when it does occur it may breed locally. Although this species can turn up almost anywhere it is scarcer in the uplands.

Scaeva species not otherwise covered:

S. albomaculata –
Vagrant from southern Europe with two known records from the south coast of England in 1938 & 1949.

S. dignota –
A recent addition to the British list that bears a very close resemblance to *S. selenitica*.

S. mecogramma –
A single record from "Aniston, Lothian" in 1905. This seems likely to have been an accidental import.

Eupeodes

9 British species (3 illustrated)

Amongst the yellow-and-black marked hoverfly genera, *Eupeodes* can be recognised by the edges of the abdomen, which from the abdomen segment T3 onwards have no yellow hairs, only short black ones. Many of the species are very challenging to identify and there is a good deal of variation in the colour, shape and extent of markings. This is associated with the temperatures at which the larvae developed. Larvae feed on conifer aphids, but those of the two commonest species, *E. corollae* and *E. luniger*, attack a wide range of ground-layer aphids.

Eupeodes luniger

Widespread

Wing length: 6·5–10 mm

Identification: A highly variable species in which the markings and colouration are strongly affected by the temperature at which the larvae develop (see *page 30*). It can also vary considerably in size. Females have a distinctive Y-shaped mark on the frons, formed by dusting. Amongst male *Eupeodes* with lunulate spots, the angle at which the eyes meet at the top of the head is less than 90° (more than 90° in other species), but this is difficult to assess accurately and requires considerable care.

Similar species: Some *Eupeodes latifasciatus* (males), *E. corollae* (both *p. 126*) and perhaps *E. lundbecki* (not illustrated). Large specimens may possibly also be confused with *Scaeva selenitica* (*p. 122*).

Observation tips: The distribution and migratory habits of this species are very similar to *E. corollae*, although it seems to have a more permanent resident population. It is likely to be one of the first species encountered in the Spring, often emerging in small numbers as early as late January or February. Peak abundance usually occurs in late Summer and it may persist through the Autumn until the first frosts.

In ♂ *Eupeodes luniger* the angle where the eyes meet at the top of the frons is < 90°.

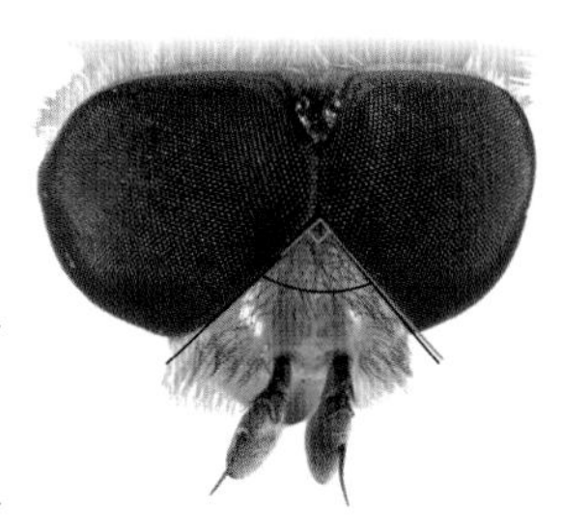

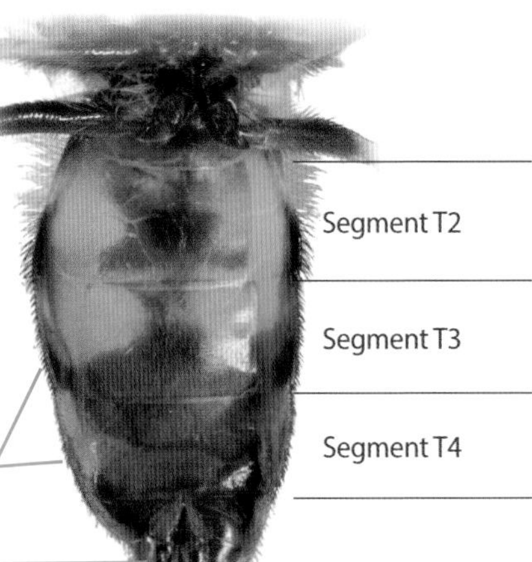

Underside of *Eupeodes corollae* abdomen showing the dark hairs on the edge of abdomen segments T3 and T4 and large male genital capsule.

♀ – *Eupeodes luniger* × **6** – ♂

Eupeodes species not otherwise covered:

Eupeodes is a particularly difficult genus, partly because the taxonomy is still in a state of flux, but also because they show a great deal of variation in colour and markings depending on the temperature at which larvae developed. Consequently, only the three commonest species have been covered here in detail.

E. lapponicus – A rather scarce species with a few records per year, usually near the south or south-west coast. Stubbs & Falk (2002) refer to *Eupeodes* 'species A': the few records are mostly inland and it may be a temperature related colour variety of the same species.

E. nielseni [NS] *E. nitens* [NS] – Scarce species that are difficult to identify. *E. nielseni* is mostly recorded from Scotland; *E. nitens* in southern England, north to the Midlands and Wales.

E. bucculatus *E. goeldlini* – *E. goeldlini* is a recently recognised split from *E. bucculatus* (probably what is referred to as 'species B' in Stubbs & Falk, 2002). These are very difficult species to identify.

E. lundbecki – On a wider European scale, *Eupeodes* can be difficult to separate from *Scaeva* and *E. lundbecki* looks more like a *Scaeva* species. It is a vagrant that has been recorded about 10 times around the British coast.

Eupeodes corollae 

Widespread

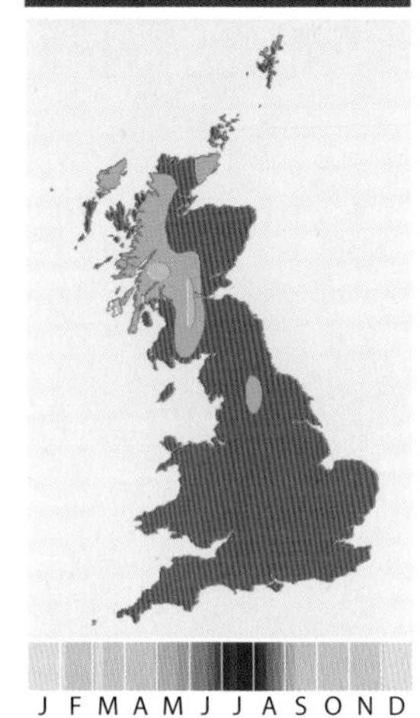

Wing length: 5–8·25 mm

Identification: An extremely variable, sexually dimorphic species in which the yellow markings are quite broad where they reach the edge of the abdomen segments. These markings occupy 50% or more of the margin in males (25% or less in other species) and this is the only species of *Eupeodes* where they reach the margins in females. Males also have an unusually large and obvious genital capsule (illustrated on *page 124*).

Similar species: Several *Eupeodes* species with lunulate markings on the abdomen may cause confusion, including *E. luniger* (*p. 124*), *E. latifasciatus* (some males) and *E. lundbecki* (not illustrated). Female *E. corollae* are not infrequently misidentified as *Parasyrphus punctulatus* (*p. 146*)

Observation tips: A migratory species whose numbers peak in mid-Summer. As a consequence, it is likely to be found almost anywhere, although its frequency declines northwards. A very common visitor to garden flowers.

Eupeodes latifasciatus

Widespread

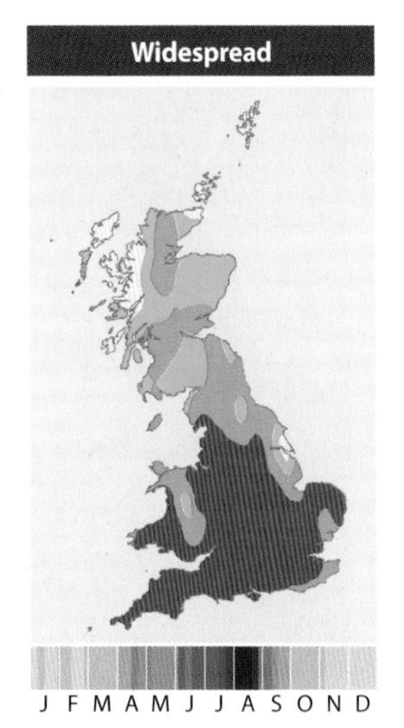

Wing length: 6·5–8·5 mm

Identification: Females are more straightforward to identify from other members of the genus than males, as the markings on the abdomen form bands and there are no dust spots on the distinctive half black, half yellow frons. Males are very variable and their abdominal markings range from bands through to lunulate spots. The most reliable character requires a microscope: the 2nd basal cell of the wing is completely covered with microtrichia (75% or less in other species).

Similar species: Some *Eupeodes luniger* (*p. 124*) and potentially *E. corollae* and *E. lundbecki* (not illustrated). *Syrphus* species (*pp. 150–153*) are frequently misidentified as this species.

Observation tips: A widespread species which occurs in a variety of situations, perhaps favouring damper woodland edges and meadows. Abundance varies considerably from year to year, suggesting that it is at least a partial migrant.

♂ – *Eupeodes corollae* × **6** – ♀

♂ – *Eupeodes latifasciatus* × **6** – ♀

Melangyna 9 British species (6 illustrated)

This is a challenging genus with several species which are very difficult to identify and others that are exceedingly rare (*M. barbifrons* and *M. ericarum*). Amongst the yellow-and-black marked genera with at least partially yellow faces, this genus is relatively small and narrow with mainly black legs. However, some species (especially *M. quadrimaculata*) can have completely black faces, which can be particularly confusing when recognising this genus. The larvae, where known, feed on a variety of aphids.

Melangyna compositarum / labiatarum 

Frequent

Wing length: 6·25–8·75 mm

Identification: There is some debate about whether these are really separate species. They have very little black on the scutellum and slightly hairy eyes. Can be confused with *M. arctica*, which has mainly black hairs on the thorax (rather than pale hairs), and *M. umbellatarum*, which has a shiny thorax (rather than dull and dusted).

Similar species: *Melangyna arctica* (not illustrated), *M. ericarum* (not illustrated), *M. lasiophthalma* and *M. umbellatarum* (both *p. 130*).

Observation tips: Occurs in woodland from May onwards, but most abundant from mid-June to mid-August. They are flower visitors, especially to umbellifers (not infrequently in the company of *M. umbellatarum*). Males occasionally form small hovering swarms.

Melangyna scutellar markings

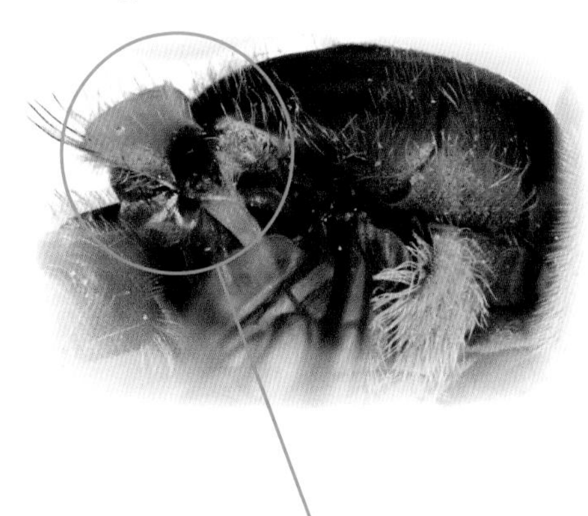

M. compositarum / labiatarum and *M. umbellatarum*:

Scutellum with black spot confined to base.

M. lasiophthalma (also *M. arctica* and *M. ericarum* – neither illustrated):

Scutellum with black spot that extends into a dark rim around the tip of the scutellum.

Melangyna compositarum / labiatarum ♀ ×**6**

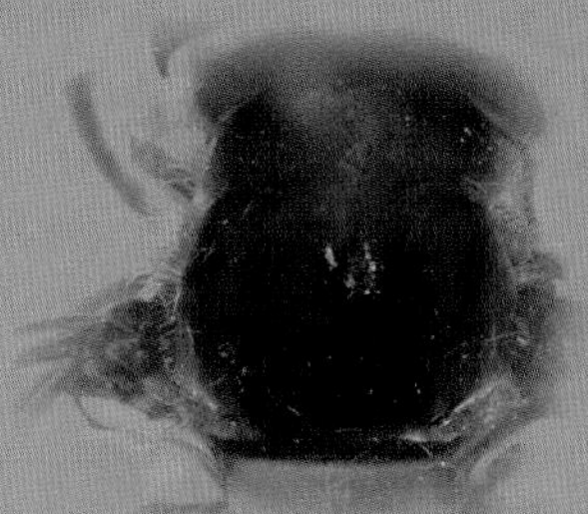

M. compositarum / labiatarum:
Thorax dull and dusted

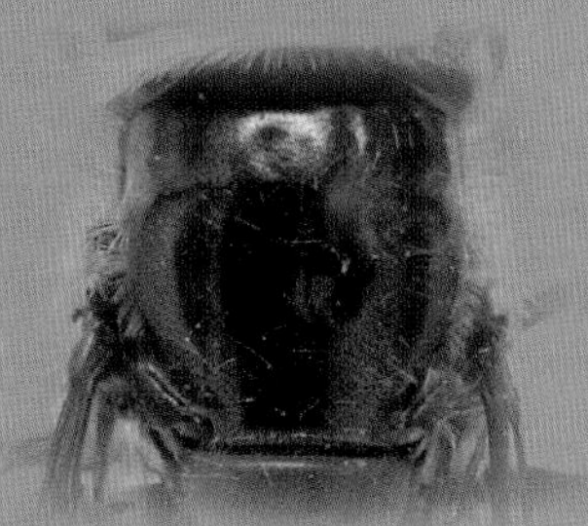

M. umbellatarum:
Thorax shiny

Melangyna species not otherwise covered:

M. arctica – A scarce species of the north and west that seems to prefer conifer woods, although it can sometimes be found in other habitats.

M. barbifrons [NT] – An extremely scarce, although widespread, woodland species that flies very early in the spring. Only 14 known records since 1980.

M. ericarum [VU] – An extremely rare species of the Caledonian Pine forest of central Scotland. Only two known records since 1980.

Melangyna lasiophthalma

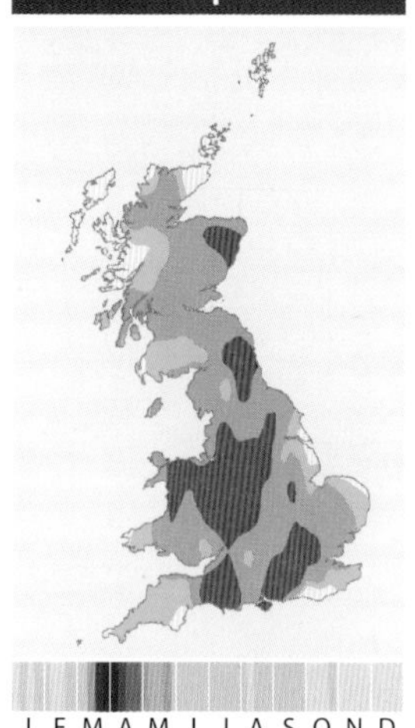

Wing length: 7–9·25 mm

Identification: The fact that this species flies so early in the year is a significant help in identification. Males have pale thoracic hairs, strong black markings on the scutellum (illustrated on *page 128*) and somewhat reduced spots on abdomen segment T2. Females have extensive black on the scutellum and narrow dusting on the frons (narrower than in *M. arctica* or *M. ericarum*).

Similar species: *Melangyna arctica* (not illustrated), *M. compositarum / labiatarum* (*p. 128*) and *M. ericarum* (not illustrated).

Observation tips: An early Spring species found visiting flowering willows, Blackthorn and other shrubs, but generally flying quite high. Since it is often found with *Parasyrphus punctulatus* (*p. 146*) and *M. quadrimaculata* (*p. 132*), *Melangyna*-like hoverflies cannot automatically be assumed to be this species!

Melangyna umbellatarum

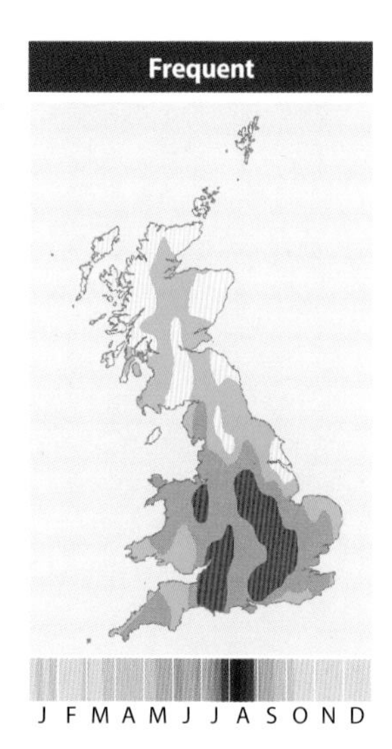

Wing length: 6·5–8·75 mm

Identification: A moderately distinctive species on two accounts: the thorax is strongly shining (see *page 129*), unlike the dull and dusted thorax of *M. compositarum / labiatarum*, with which it is most likely to be confused; and the abdominal markings are paler than is typical for the genus, often creamy or almost white.

Similar species: *Melangyna arctica* (not illustrated), *M. compositarum / labiatarum* (*p. 128*) and *M. ericarum*(not illustrated).

Observation tips: Most frequently found at Hogweed and other umbellifers in woodland rides or in scrubby grassland. Mainly encountered in mid-Summer, when it can sometimes be numerous. Predominantly a southern species, but occurs north into the highlands of Scotland and the Spey valley.

Melangyna lasiophthalma ♂ ×**6**

Melangyna umbellatarum ♂ ×**6**

Melangyna cincta

Wing length: 6·25–8·75 mm

Identification: One of the more obvious members of the genus, with pointed, triangular markings on abdomen segment T2 and complete bands on the other segments. Most likely to be confused with *Meliscaeva cinctella*, in which the markings on the abdomen segment T2 are rounded.

Similar species: *Meliscaeva cinctella* (*p. 136*).

Observation tips: A woodland species, which although widespread, is rarely abundant and usually found visiting flowers such as Hogweed. It is less common in the north. The larvae are known to feed on a variety of tree-dwelling aphids.

Frequent

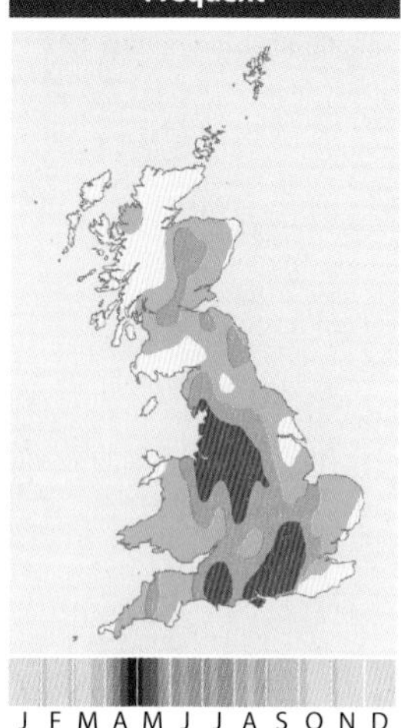

Melangyna quadrimaculata

Wing length: 7·25–9 mm

Identification: A tricky species to recognise because the face is often very dark, suggesting that it is not a member of the tribe Syrphini; and also because females lack abdominal markings. Both males and females are most likely to be confused with *M. barbifrons*, though males of this species have dark thoracic and scutellar hairs (pale in *M. quadrimaculata*), and females have abdominal markings.

Similar species: Apart from *Melangyna barbifrons* (not illustrated), the most likely confusion is with *Cheilosia* species that have hairs on their faces (particularly those of the VARIABILIS group (*p. 164*)), though the presence of a zygoma should eliminate this genus.

Observation tips: A very early Spring species, with a peak flight period from the end of March to early April. Visits willow flowers in woodland rides (especially in coniferous woodlands), but often flying at considerable height. In many years, poor weather can lead to under-recording and it is probably more frequent than records suggest.

Local

Melangyna cincta ♂ ×**6**

♂ – *Melangyna quadrimaculata* ×**6** – ♀

Meligramma

3 British species (2 illustrated)

Narrow-bodied hoverflies that are similar to *Melangyna* (some authors include them in that genus). They tend to be small and have a yellow scutellum amd a yellow face that lacks any dark central 'knob'. They can look rather like *Platycheirus* (*pp. 78–91*) although that genus has a black face. *Melangyna* (*pp. 128–133*) is also similar but the face is only partially yellow and has a darkened central 'knob'. The larvae feed on aphids on trees and shrubs; *M. guttatum* has been reared from aphids on Sycamore.

Meligramma guttatum

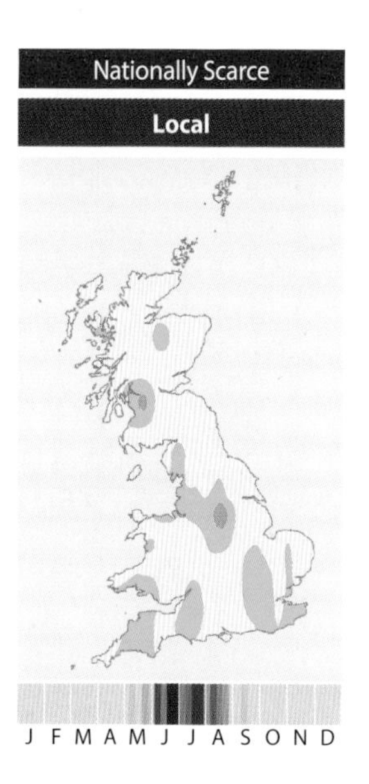

Wing length: 5·25–7 mm

Identification: A small, narrow-bodied hoverfly with yellowish or whitish spots on the abdomen and a yellow face without a black facial 'knob'. The markings on abdomen segment T2 are often very small and indistinctly triangular and those on segments T3 and T4 are rounded, making it distinctly different from other species in the genus. Tends to have yellow spots at the back of the thorax, just in front of the scutellum.

Similar species: Possibly some *Platycheirus* (*pp. 78–91*), although confusion seems unlikely.

Observation tips: A woodland species which occurs sporadically in low numbers across the country, often along riverbanks and in damp places in southern England, and in Sycamore woods further north. Adults visit a range of flowers including Hogweed and Angelica, mainly in June and July.

Meligramma triangulifera

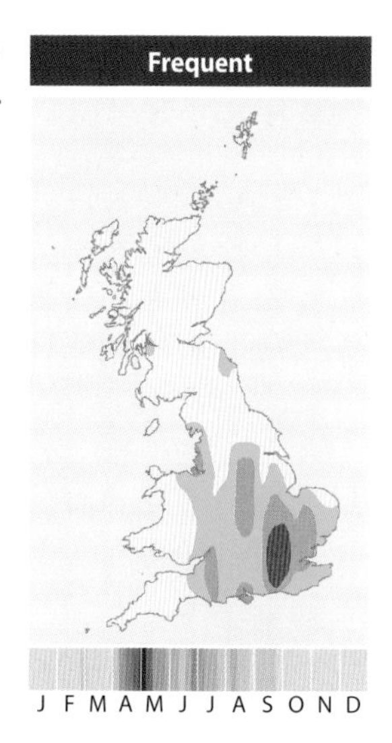

Wing length: 5–8 mm

Identification: A narrow-bodied hoverfly with a yellow face and with yellow spots on the abdomen. The markings on abdomen segment T2 are often very small and are approximately triangular. It is a somewhat cryptic species that might be overlooked among other narrow-bodied hoverflies.

Similar species: Possibly some *Platycheirus* (*pp. 78–91*), especially more yellowish-marked *P. albimanus* males, but confusion will be readily dispelled under magnification.

Observation tips: A species of woodland edge, hedgerows and scrub which is most frequently found in Spring, often in late April or May. Although it does visit flowers such as those of Field Maple, it is more frequently seen basking on sunlit leaves.

Meligramma guttatum ♀ ×**6**

Meligramma species not otherwise covered:

M. euchromum [NS] – A rather scarce southern species with most records from Surrey and Hampshire. Looks similar to some *Melangyna* and often reported erroneously.

Meligramma trianguliferum ♂ ×**6**

Meliscaeva
2 British species (both illustrated)

Small, elongate, black-and-yellow hoverflies with yellow faces and somewhat elongated wings. They are woodland species which occur in highly variable numbers, especially *M. auricollis* which can, seemingly, barely occur in some years. The larvae feed on aphids on various shrubs and trees.

Meliscaeva auricollis

Widespread

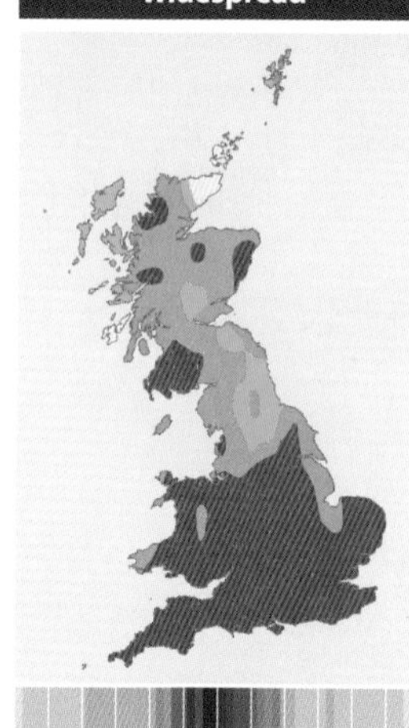

J F M A M J J A S O N D

Wing length: 6–9·5 mm 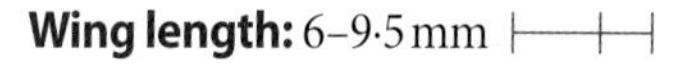

Identification: An extremely variable species in which the markings are strongly influenced by the temperature at which the larvae develop (see *page 30*). In Spring, if the larvae have developed in cold conditions, the adults are often darker and less well-marked than in later generations whose larvae developed in warmer conditions. The elliptical markings on abdomen segment T2 are a useful feature. The oblique hind margins of the spots on abdomen segments T3 and T4 are distinctive, but vary hugely and sometimes fuse to form bands which can make it hard to separate from *M. cinctella*. In these cases, features of the face (illustrated *opposite*) can be useful.

Similar species: *Meliscaeva cinctella*. Members of the *Platycheirus* PELTATUS group (*p. 84*), and especially weaker-marked *P. scutatus*, which has a black face. *Melangyna cincta* (*p. 132*) is similar in general shape but has sharply triangular yellow markings on abdomen segment T2.

Observation tips: Has a very long flight period, but numbers peak in late Summer. Predominantly a woodland edge species which occurs throughout Britain.

Meliscaeva cinctella

Widespread

Wing length: 7–9·75 mm

Identification: A relatively long-and-narrow species which can only be confused with a small number of other small, banded hoverflies such as *Parasyrphus*. The combination of broad, blunt-ended markings on the abdomen segment T2, combined with bands on abdomen segments T3 and T4, is distinctive.

Similar species: *Meliscaeva auricollis. Melangyna cincta* (*p. 132*) is similar in general shape but has sharply triangular yellow markings on abdomen segment T2. *Parasyrphus* (*pp. 146–149*).

Observation tips: A widespread woodland species which visits a wide range of flowers such as Bramble, ragworts and umbellifers. It often occurs in considerable numbers in mid-Summer.

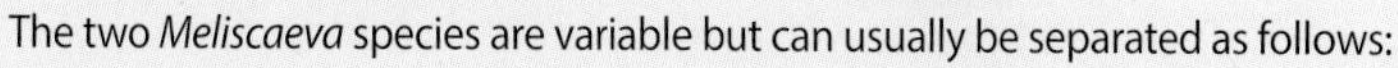

The two *Meliscaeva* species are variable but can usually be separated as follows:

***M. auricollis*:** ABDOMEN: markings on segment T2 **elliptical**
***M. cinctella*:** ABDOMEN: markings on segment T2 **broad with blunt ends**

Episyrphus

1 British species (illustrated)

Britain's commonest hoverfly, commonly known as the Marmalade Fly. This is an extremely variable species whose background colour is highly influenced by the temperature at which the larvae developed (see *page 30*). Larvae in hot conditions produce adults with more orange markings (sometimes almost lacking any black markings), whilst those that develop in cooler conditions produce darker adults (sometimes completely black). The larvae feed on a wide range of aphid species and can be very numerous in agricultural crops, for example feeding on cereal aphids and Cabbage Aphid.

Episyrphus balteatus

Widespread

Wing length: 6–10·25 mm

Identification: Each abdomen segment has two dark bands separated by two orange bands. This is a feature that is not shared by any other British hoverfly species.

Similar species: None.

Observation tips: Ubiquitous and very common. Numbers tend to rise through the Spring and there is often a marked peak in abundance in late July, but adults can occur at any time of year and, as they hibernate, can even be found on a sunny day in mid-Winter. Mass immigrations from continental Europe occur and sometimes lead to reports in the press of 'plagues of wasps'.

May be seen in some numbers in mid-Summer.

Typical colouration

▲ ♂ – *Episyrphus balteatus* × **6** – ♀ ▼

A rather dark coloured individual photographed in March

Epistrophe

6 British species (5 illustrated)

Black-and-yellow banded hoverflies with mainly yellow faces that fly in the Spring and early Summer (except *E. grossulariae*, which flies mid- to late Summer). Apart from *E. eligans*, which is quite distinctively marked (although very variable), **they are tricky to separate from other yellow-banded genera** such as *Syrphus* (*pp. 150–153*) and *Parasyrphus* (*pp. 146–149*). The genus *Syrphus* is distinguished by the presence of hairs standing upright on the upper surface of its squamae (as opposed to long hairs around the rim which are present in other yellow-banded Syrphines) – see *opposite*. Yellow-banded *Epistrophe* lack these upstanding hairs. There are several additional European species which could turn up in Britain and the genus includes some groups of species where identification is challenging. The larvae feed on tree-dwelling aphids and *E. grossulariae*, in particular, feeds mainly on Sycamore Aphid.

Epistrophe nitidicollis

Frequent

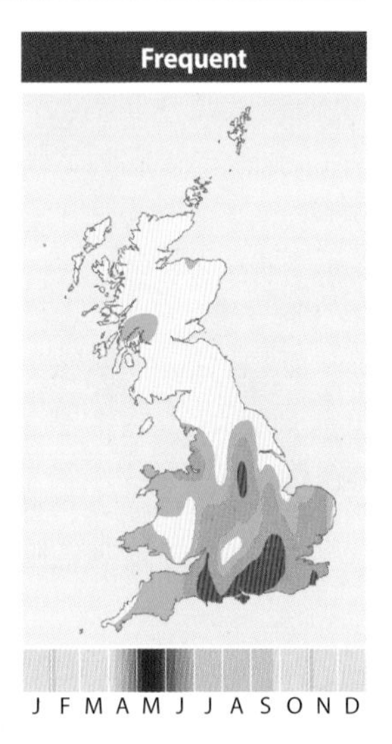

Wing length: 8–11·25 mm

Identification: This is the commonest of the banded *Epistrophe* with orange antennae, and can be separated from others in this group by its black scutellar hairs. Care is needed with this character as there are several similar species; in particular, it might be confused with *Parasyrphus nigritarsis*, except that this species has tarsi that are black; yellow in *E. nitidicollis*.

Similar species: Apart from confusion with the Nationally Scarce *Epistrophe melanostoma* (not illustrated), *E. ochrostoma* (not illustrated), *Megasyrphus erraticus* (*p. 144*) and *Parasyrphus nigritarsis* (*p. 148*), it is most likely to be overlooked as a *Syrphus* (*pp. 150–153*).

Observation tips: A woodland species which visits a wide range of plants along rides, although it seems to favour Garlic Mustard. Most records are in May and June and anything later should be treated with caution and checked for other species.

Epistrophe species not otherwise covered:

E. melanostoma [NS]	– A relatively recent addition to the British fauna that is difficult to identify. Requires careful comparison with known specimens to distinguish from *E. nitidicollis*. Few records from the south east of England.
E. ochrostoma [DD]	– A single, somewhat doubtful record from North Wales in 1990. Unfortunately, the specimen has been lost so it cannot be checked!

Epistrophe nitidicollis and E. melanostoma scutella.

Epistrophe nitidicollis ♀ ×**6**

Telling yellow-banded species of the Syrphini from *Syrphus*

Other yellow-banded species
Upper surface of squamae without hairs

Syrphus
Upstanding hairs on the upper surface of squamae

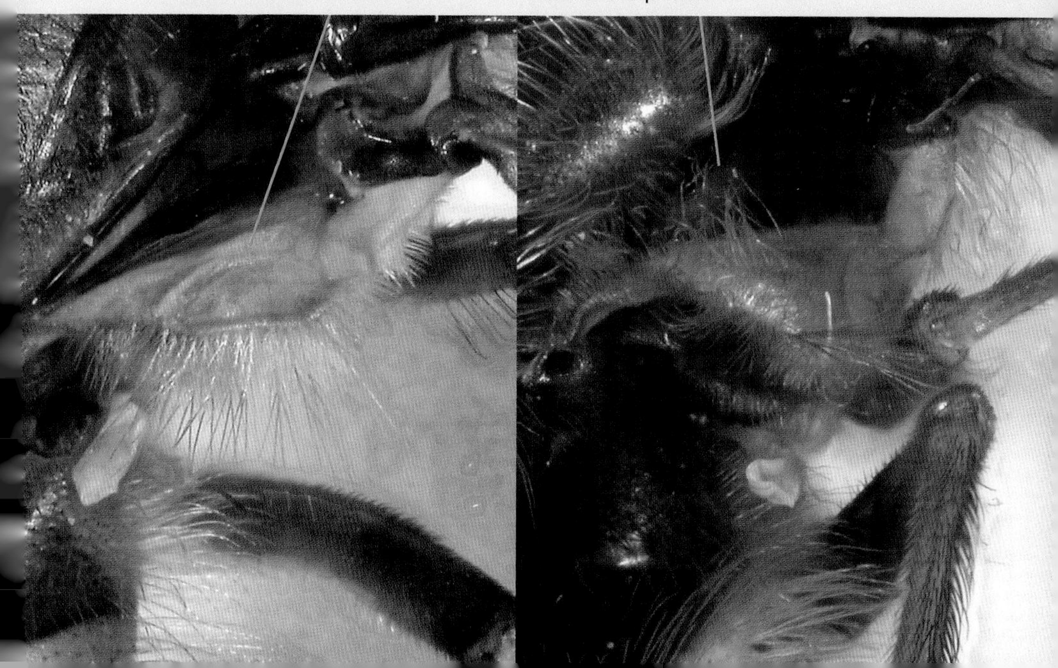

Epistrophe diaphana

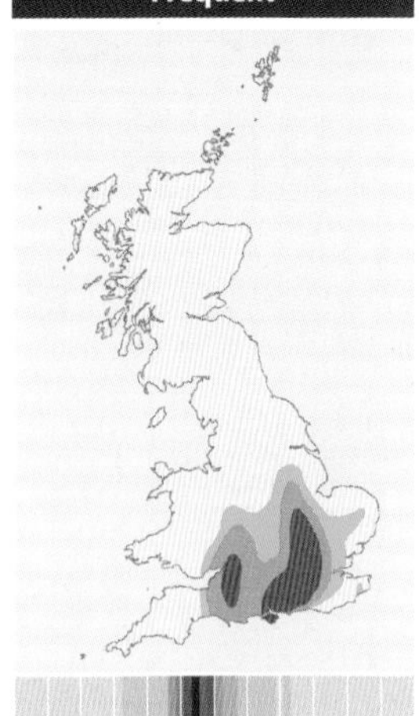

Wing length: 7·25–9·75 mm

Identification: A banded *Epistrophe* with black antennae is either this species or *E. grossulariae*, although *E. diaphana* has yellow front femora (partly black in *E. grossulariae*). Additional features include the shape of the margins of the bands on the abdomen, which are turned forwards (straight in *E. grossulariae*). It may be picked out amongst a plethora of *Syrphus* because it tends to be more orange, but other characters must be checked.

Similar species: Photographs of *Syrphus* (*pp. 150–153*) are often mistaken for this species. Any of the common *Syrphus* can be mistaken for it, although in the field *Epistrophe diaphana* has a much more orange abdominal colouration. Some banded *Eupeodes*, especially *E. latifasciatus* (*p. 126*) and *E. bucculatus* (not illustrated) may cause confusion but this genus can be recognised by the diagnostic abdominal edge hairs (see *p. 124*).

Observation tips: A woodland and scrub-edge species often seen visiting Hogweed and other umbellifers. Although a southern species, it has become more frequent in the Midlands in recent years and may be expanding its range northwards.

Epistrophe grossulariae

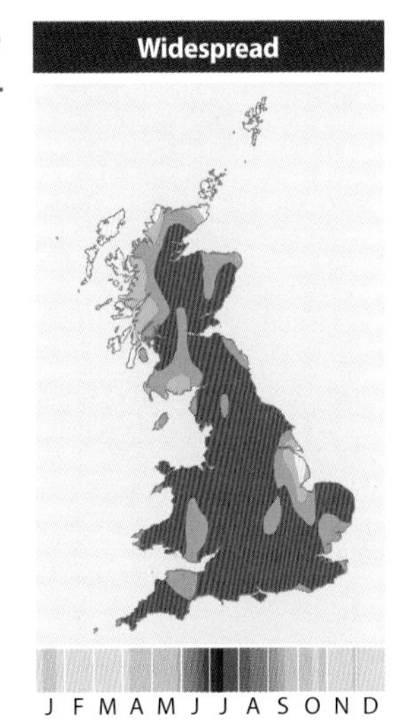

Wing length: 9–12·25 mm

Identification: Most likely to be confused with *Syrphus* (but does not have hairs on the upper surface of the squamae) amongst which it may be picked out by its more orange colour. Once it is recognised as an *Epistrophe*, the dark antennae, black bases to the front femora and the shape of the bands on abdomen segments T3 and T4 (which are not turned forwards as in *E. diaphana* but run straight to the edges or turn slightly backwards) should distinguish it.

Similar species: Most likely to be overlooked as a *Syrphus* (*pp. 150–153*) but may be confused with the Nationally Scarce *Epistrophe melanostoma* (not illustrated), *E. nitidicollis* (*p. 140*), *E. ochrostoma* (not illustrated), *Megasyrphus erraticus* (*p. 144*), which has a dark, shiny thorax and *Parasyrphus nigritarsis* (*p. 148*), which has black tarsi.

Observation tips: A woodland edge species which occurs widely across mainland Britain. It appears to be more frequent in northern and western Britain, which suggests that it favours damper environments.

Epistrophe diaphana ♀ ×**6**

Epistrophe grossulariae ♀ ×**6**

Epistrophe eligans

Wing length: 6·25–9·5 mm

Identification: A shining, brassy-coloured hoverfly with yellow markings mainly on abdomen segment T2 and absent or much reduced on the other segments, but very variable.

Similar species: Unlikely to be mistaken for anything else, although some small *Eristalis* (*pp. 202–204*) may look superficially similar (but these have a loop in veins R_{4+5}).

Observation tips: Primarily a Spring species which visits a wide range of flowers. The larvae are associated with aphids on trees and shrubs, favouring Elder, Sycamore and fruit trees. Males often form swarms, competing for sunlit spots in woodland rides or around a particular bush. It is one of a small number of species that have been found to be highly responsive to warmer Springs and it now emerges much earlier than it did even 20 years ago. Occurs into northern England and coastal and lowland Scotland, but much less common farther north.

Megasyrphus

1 British species (illustrated)

This is a genus with a single species which has distinct black-and-yellow bands on the abdomen. It is likely to be overlooked as a *Syrphus* (*pp. 150–153*). Closer inspection reveals a distinct dip in the vein R_{4+5}, and a broad black stripe across the undersides of the abdominal segments. The larvae feed on a variety of aphids, especially those feeding on broadleaved trees such as Alder and Ash.

Megasyrphus erraticus

Wing length: 10·75–12 mm

Identification: Larger than most other hoverflies with banded abdomens and tends to look rather more orange and shiny in comparison. Unlike *Syrphus*, it has a black facial stripe and the thorax is darker and more shiny (the thorax of *Syrphus* are olive-coloured and somewhat dull).

Similar species: It is most likely to be confused with *Syrphus* (*pp. 150–153*) but might also be confused with banded members of *Epistrophe* (*pp. 140–145*), which have a duller, more olive-coloured thorax and *Parasyrphus nigritarsis* (*p. 148*), which has black tarsi in both sexes and a black hind femur in the male.

Observation tips: This is primarily a northern and western species which is mainly found in coniferous plantations. It will visit a wide variety of flowers, especially yellow composites. Rides or clearings with a stream are good places to look.

See *Identifying wasp and bee mimics* on *pages 64–66*.

♀ – *Epistrophe eligans* × **6** – ♂

Male with rather reduced markings on the abdomen

Megasyrphus: SYRPHINI

Megasyrphus erraticus ♂ × **6**

Parasyrphus

6 British species (3 illustrated)

Black-and-yellow banded hoverflies with mainly dark hind legs. Most likely to be confused with *Syrphus* (*p. 150*), except that the squamae do not have hairs on the upper surface (see *page 141*). Most are associated with coniferous woodlands and many are difficult to identify because the degree of yellow marking on the knees of the hind legs is both variable in extent and intensity. Larvae feed on aphids on trees, especially conifer aphids.

Parasyrphus punctulatus

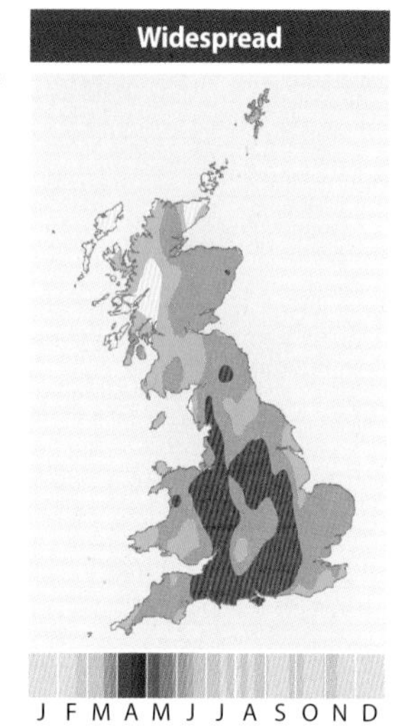

Wing length: 5·5–7·75 mm

Identification: The shape of the abdominal markings differ between males and females, making this species difficult to identify in the field. It is unusual amongst *Parasyrphus* because the abdominal markings are paired spots rather than bars. However, the combination of this feature and the darkened legs can easily lead to confusion with *Melangyna*.

Similar species: *Melangyna* species (*pp. 128–133*) and *Meligramma euchromum* (not illustrated but see *page 135*). Photographs of both *Eupeodes luniger* (*p. 124*) and female *E. corollae* (*p. 126*) have sometimes been labelled erroneously as *P. punctulatus*.

Observation tips: Commonest in the early Spring when it visits the blossom of trees such as willows and Bird Cherry (often flying high up), but can be found through to mid-Summer in smaller numbers. Most common in woodland, particularly coniferous woodlands.

Parasyrphus species not otherwise covered:

P. annulatus – A comparatively small *Parasyrphus* with the hind femur yellow at the base (illustrated *page 149*). Widespread but very local, usually in conifer woods.

P. lineola – This species has black antennae and entirely black hind legs (sometimes a little yellow around the knee). Widespread and not uncommon in conifer woods. It can be overlooked amongst *Meliscaeva cinctella*.

P. malinellus – Rather similar to *P. lineola*, with the hind femur and tibia largely black usually with a somewhat more extensively yellow knee (illustrated *page 149*). Widespread in conifer woods, but most abundant in Scotland.

♀ – *Parasyrphus punctulatus* × **6** – ♂

Parasyrphus abdomens

P. punctulatus
'golf-club'-shaped
spots on segments
T3 and T4

All other *Parasyrphus*
'moustache'-shaped
bands on segments
T3 and T4

Parasyrphus nigritarsis

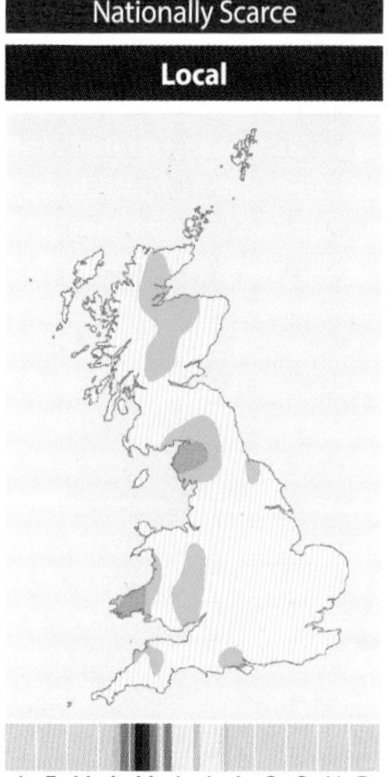

Wing length: 8·25–9·25 mm

Identification: Looks very like a *Syrphus* in general appearance. Although this species is very difficult to identify with certainty, useful pointers include the absence of hairs on the upper surface of the squamae, a distinct moustache-shaped band on the abdomen segments T3 and T4, a completely yellow face and black tarsi.

Similar species: *Epistrophe* (*pp. 140–145*), *Megasyrphus erraticus* (*p. 144*), *Syrphus* (*pp. 150–153*) and some banded *Eupeodes* (*pp. 124–127*).

Observation tips: A northern and western species which was once regarded as a rarity. The larvae feed on leaf-beetle larvae on Alder and willows along rivers and streams, and adults can be found in such places basking on sunlit leaves.

Parasyrphus vittiger

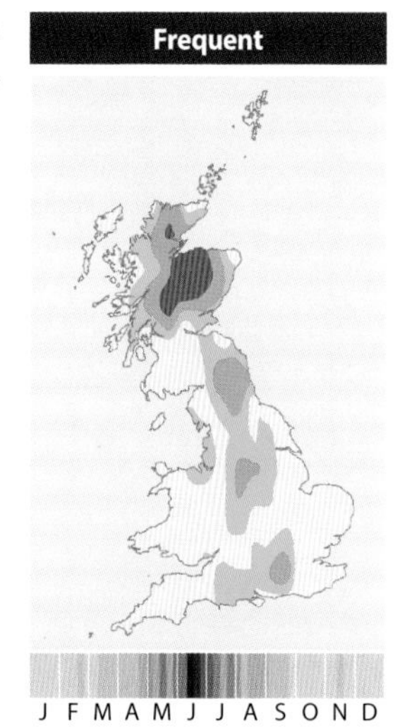

Wing length: 6·25–8·75 mm

Identification: This is one of four very similar species (with *P. annulatus*, *P. malinellus* and *P. lineola*), which are separated on the extent of yellow markings on the hind leg (see *opposite*).

Similar species: *Parasyrphus annulatus*, *P. lineola* and *P. malinellus*, and some banded *Eupeodes* (*pp. 124–127*).

Observation tips: Conifer plantations and heathland with conifers. The adults visit yellow flowers both along rides and on the open heath. Mainly a northern species, but also widespread on the heathlands of southern England. Numbers tend to peak in June, but on southern heathland it can also be quite abundant in late Summer.

Parasyrphus legs

P. nigritarsis
tarsi black

hind leg extensively yellow (even more so in female)

Parasyrphus nigritarsis ♀ ×**6**

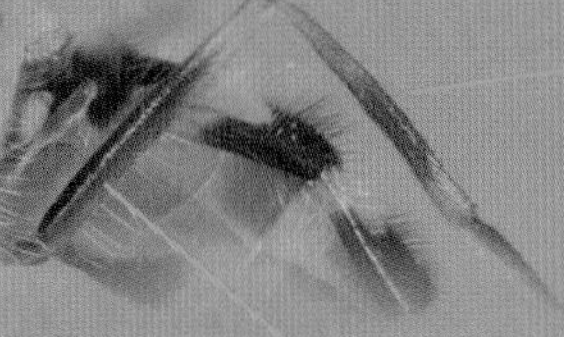

P. vittiger
tibia with dark ring

hind femur black at base

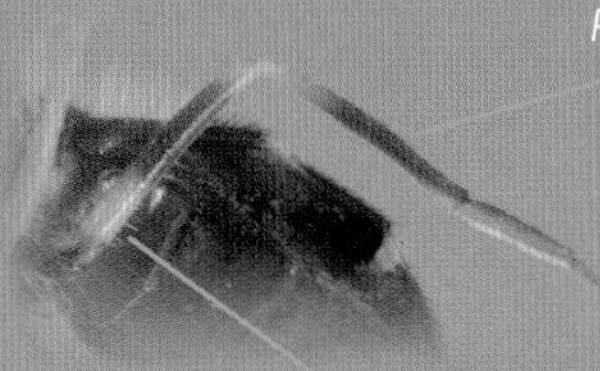

P. annulatus
tibia with more extensive dark ring

hind femur yellow at base

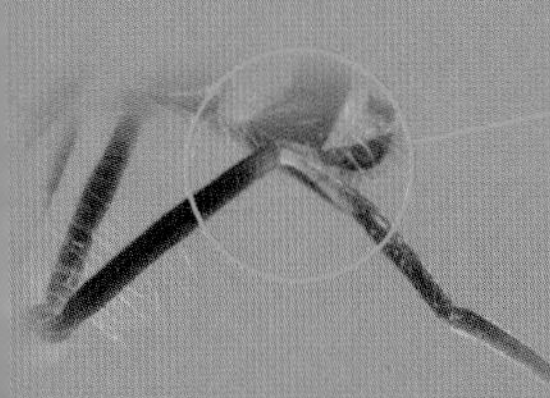

P. malinellus
hind leg almost all black except for broadly yellow 'knee'

P. lineola (not illustrated) is even blacker and the yellow at the 'knee' is more restricted

Parasyrphus vittiger ♂ ×**6**

Syrphus

5 British species (3 illustrated)

Yellow-and-black banded hoverflies with a yellow face and the dusted thorax having a dull bronzy greenish appearance. The defining feature of the genus, the presence of hairs on the upper surface of the squamae (see *page 141*), requires careful examination under good magnification. Three species (*S. ribesii*, *S. torvus* and *S. vitripennis*) are amongst the commonest yellow-and-black banded flower-visiting hoverflies, but are difficult to separate. There are two other British species: *S. rectus*, which may not be valid as a European species; and *S. nitidifrons*, which has only recently been added to the British list. Larvae feed on a wide variety of aphids.

Syrphus ribesii

Wing length: 7·25–11·5 mm

Identification: Females are easily recognised by their almost completely yellow hind femora. Males can be separated from *S. torvus* by their bare eyes, but separation from *S. vitripennis* (*p. 152*) is difficult. Under a microscope, the 2nd basal cell of *S. ribesii* is completely covered with microtrichia (see table *below*).

Similar species: All banded *Syrphus*, some *Epistrophe* (*pp. 140–145*) (especially *E. diaphana* and *E. grossulariae*), *Megasyrphus erraticus* (*p. 144*) and *Parasyrphus* (*pp. 146–149*).

Observation tips: Occurs from early Spring to Autumn in most lowland localities throughout Britain, although numbers vary hugely. The males form small swarms which hover in dappled light under trees and emit much of the familiar buzz of a Spring woodland.

Identification features of *Syrphus*.

SPECIES	MALES: Eyes meet on top of head		WING (both sexes):	FEMALES: Eyes well separated on top of head
S. ribesii	EYES: bare	HIND FEMUR: yellow area at end with at least **some black hairs**	2nd basal cell entirely covered by microtrichia	HIND FEMUR: **mostly yellow**, black at extreme base only
S. vitripennis		HIND FEMUR: yellow area at end **with yellow hairs**	2nd basal cell **partially** covered by microtrichia	HIND FEMUR: extensively darkened, last 1/3 – 1/4 yellow
S. torvus	EYES: **hairy**		2nd basal cell entirely covered by microtrichia	
S. rectus	Males cannot be separated from *S. vitripennis*; female identification doubtful			
S. nitidifrons	Difficult to be sure it is a *Syrphus* because the hairs on the squamae can be sparse and inconspicuous. Most likely to be confused with *Parasyrphus punctulatus* (*p. 146*)			

Syrphus ribesii ♀ ×**6**

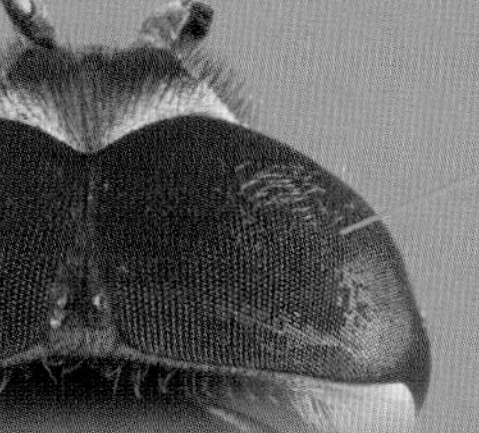

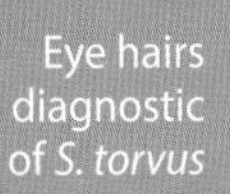

Eye hairs diagnostic of *S. torvus*

Male: eyes meet on top of head

Female: eyes well separated on top of head

♂ *S. ribesii*
femur with at least some black hairs

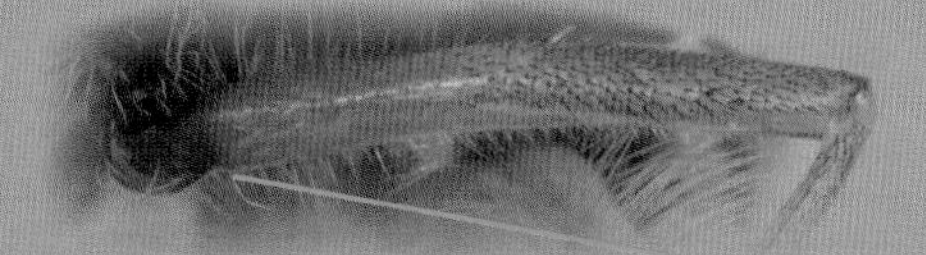

♀ *S. ribesii*
femur yellow with black at extreme base only

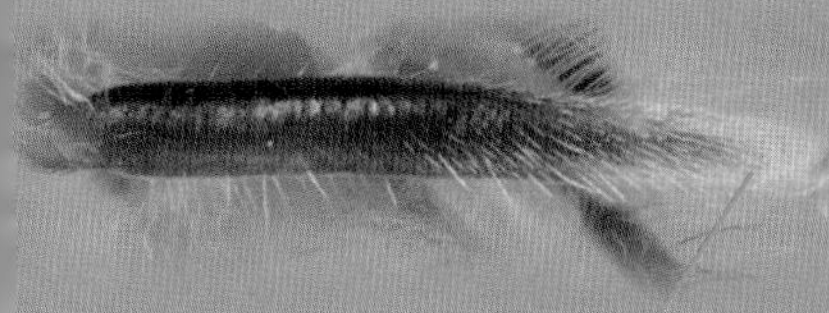

♂ *S. vitripennis*
femur with yellow hairs

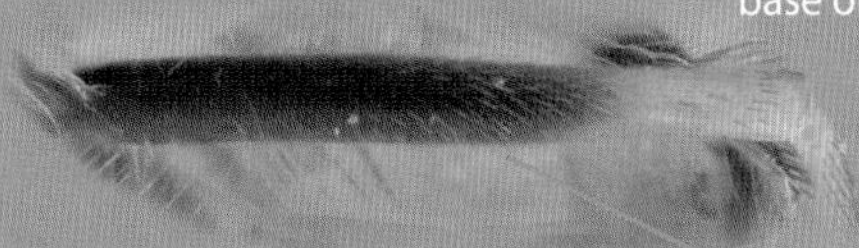

♀ *S. vitripennis / S. torvus*
hind femur extensively darkened

Syrphus torvus

Wing length: 8·5–11·75 mm

Identification: The eyes of males are covered in pale hairs which can be seen through a 20× hand lens. In females, the eye hairs are tiny, sparse and very hard to see; the best differentiating feature from female *S. vitripennis* is the 2nd basal cell of the wing, which in both sexes is completely covered with microtrichia – see *below*.

Observation tips: Widespread and common throughout lowland Britain, but more abundant in the Spring than later in the season.

Similar species: All banded *Syrphus*, some *Epistrophe* – especially *E. diaphana* and *E. grossulariae* (*pp. 140–153*), *Eupeodes latifasciatus* (*p. 126*) *Megasyrphus* (*p. 144*) and *Parasyrphus* (*pp. 146–149*).

Widespread

J F M A M J J A S O N D

Syrphus vitripennis

Wing length: 7·25–10·25 mm

Identification: Under the microscope, both sexes have only partial coverage of microtrichia on the 2nd basal cell (see *below*), which distinguishes this species from *S. ribesii* and *S. torvus*. Males are hard to separate from *S. ribesii*; females are difficult to separate from *S. torvus* (see table on *page 150*).

Observation tips: Widespread and abundant in lowland Britain, but numbers fluctuate enormously due to the fact that it is one of the species involved in mass immigration from continental Europe. Can be a common visitor to garden flowers in mid-Summer.

Widespread

J F M A M J J A S O N D

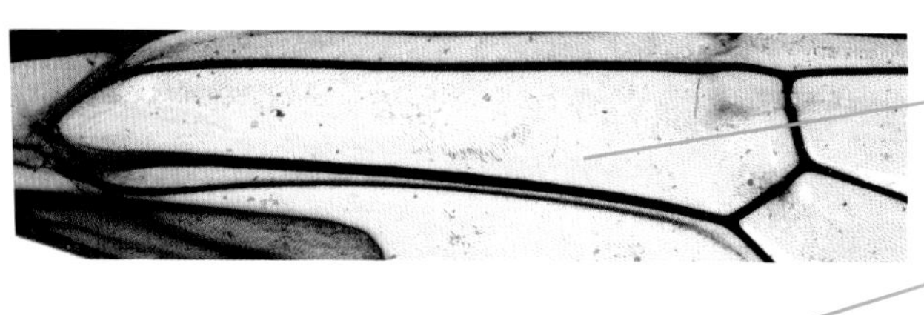

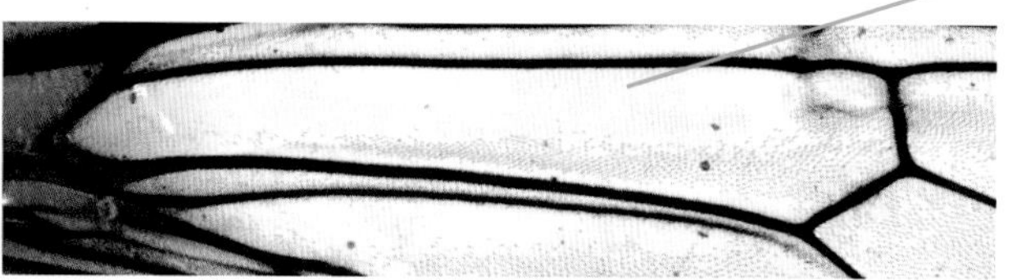

S. ribesii and *S. torvus*: entirely covered with microtrichia

S. vitripennis: partially covered with microtrichia, leaving a bare margin along the front of the cell

Wings of *S. torvus* and *S. vitripennis* showing the characteristics of the 2nd basal cell.

Syrphus torvus ♂ ×**6**

Syrphus vitripennis ♀ ×**6**

Guide to Callicerini

Represented by a single genus of charismatic hoverflies, easily recognisable by their **porrect antennae with a white-tipped terminal arista**.

Callicera

3 British species (3 illustrated)

Highly charismatic species that are rarely encountered as adults. They are more easily found by searching for larvae, which live in water-filled rot holes. However, trees with rot holes are frequently regarded as dangerous – which can lead to them being felled in urban areas or parklands. This is a particular concern in relation to *C. aurata* and *C. spinolae* (the latter being a UK Biodiversity Action Plan (UKBAP) priority species (see *page 296*)), which are most likely to occur in such situations.

Callicera rufa

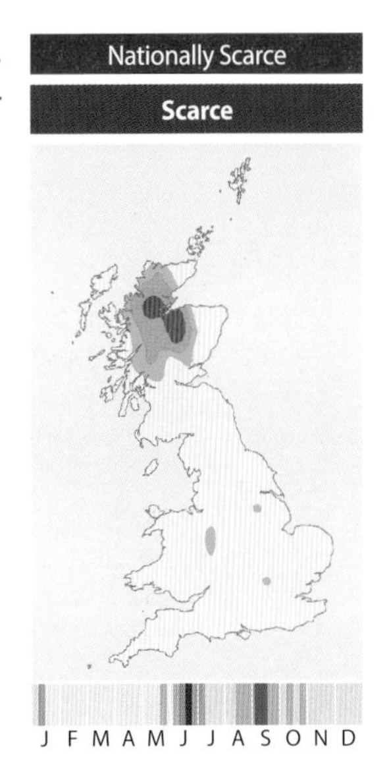

Wing length: 9·75–11·25 mm

Identification: Highly distinctive with foxy fur and yellow legs, except for the two outermost tarsal segments, which are black. Tends to be smaller than the other two *Callicera* species and the underlying body colour is black rather than bronzy.

Similar species: Only likely to be confused with other *Callicera* species.

Observation tips: A Scottish species of Caledonian pine woods, where the larvae live in rot holes in Scots Pine. In recent years it has also been found in rot holes in stumps of a variety of other conifers in felled plantations. Recent searches for larvae have shown it to be quite widespread in the highlands of Scotland and not as rare as previously thought (see *page 45*). In the last three years this species has undergone a dramatic range expansion and has been discovered in conifer plantations in Bedfordshire, Nottinghamshire and Shropshire. It is possible that it will become more widespread in England and Wales in future.

Guide to British *Callicera* species

SPECIES	FEATURES	
C. rufa	FORM: distinctive with foxy fur and base colour **black**	
C. aurata	FORM: base colour **bronzy**	LEGS: ♀ basal half of femora **black** ♂ **black** hairs between front and middle legs
C. spinolae		LEGS: ♀ basal half of femora **yellow** ♂ **yellow** hairs between front and middle legs

C. rufa larva

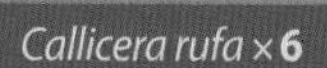

Callicera rufa × **6**

Callicera aurata

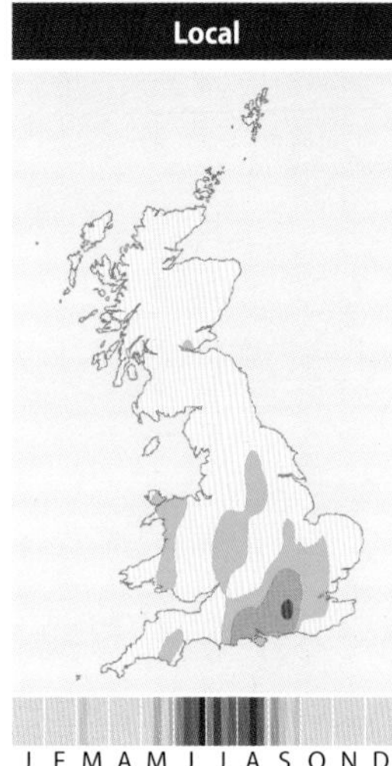

Wing length: 10–12·5 mm

Identification: Females can be separated from the other two *Callicera* species by the basal halves of their femora, which are black. Males are difficult to separate from *C. spinolae* but have black hairs between the front and middle legs (yellow in *C. spinolae*).

Similar species: Only likely to be confused with other *Callicera* species, especially *C. spinolae*.

Observation tips: Most records are from southern England, although there is a recent Scottish record. The larvae live in water-filled cavities in a variety of trees. Adults visit flowers such as Hawthorn, umbellifers and Ivy. Usually flies between June and August, but has occasionally been reported into September, when the flight period overlaps with that of *C. spinolae*.

Callicera spinolae

Wing length: 12–15 mm

Identification: Females can be separated from *C. aurata* by their entirely yellow legs. Males are difficult to separate and identification relies on the colour of hairs between the front and middle legs (tricky to see).

Similar species: Only likely to be confused with other *Callicera* species, especially *C. aurata*. It tends to fly later in the year, but care is needed because *C. aurata* has also been found occasionally in Autumn.

Observation tips: Currently known from only a few localities in East Anglia. Adults fly in Autumn and visit Ivy flowers. As a UKBAP priority species, it has been the subject of detailed studies, including extensive searches for larvae. This has shown that it can use remarkably small rot holes in a variety of tree species. There have been more records in recent years and it is possible that the species is benefiting from warmer Summers as a result of climate change.

Callicera aurata ♂ ×**6**

Callicera spinolae ♀ ×**6**

Guide to Cheilosiini

There are four genera within this tribe including our largest genus (*Cheilosia*). The humeri are hairy and the head is sufficiently separated from the thorax to make them visible in many cases. **The Cheilosiini can be readily identified by the presence of a zygoma** – a flat margin to the eye, below the antennae, which is clearly separated from the rest of the face by a fold or groove – see *opposite* and *page 160*. Most species are black but a few are more brightly coloured, *e.g. Rhingia* (*p. 178*), or are covered in coloured fur, *e.g. Cheilosia illustrata* (*p. 162*) and *C. chrysocoma* (*p. 174*).

1

a) Colourful species:

Abdomen with rectangular grey dust markings (when fresh)

Face without obvious 'nose' but lower half projecting somewhat

Antennae orange

Portevinia maculata p. 176

Beware - some *Platycheirus* (*pp. 78–91*) also have grey markings on the abdomen.

Portevinia maculata

Thorax with distinct grey stripes; abdomen brassy, wings with dark markings

Ferdinandea p. 176

Ferdinandea cuprea

Mouth edge projecting into an obvious 'beak'

Abdomen orange

Rhingia p. 178

Rhingia campestris

b) Less colourful species:

Face with distinct 'nose' 

2 *from* **1b**

a) Furry bee mimics

Cheilosia albipila p. 168
Cheilosia chrysocoma p. 174
Cheilosia grossa p. 168
Cheilosia illustrata p. 162

Cheilosia chrysocoma

b) Black hoverflies with few or no bristles

Note: this is our largest genus and includes species of very varied size and form.

Cheilosia pp. 160–175

Cheilosia bergenstammi

1a) and 1b) Cheilosiini faces

The presence of the zygoma is unique to the Cheilosiini among British hoverflies.

RIGHT: *Cheilosia vulpina* showing the zygoma and distinct 'nose'.

BELOW RIGHT: *Portevinia maculata* illustrating the zygoma, the lack of an obvious 'nose' and the slight projection of the lower half.

BELOW: *Rhingia rostrata* showing the more extreme 'beaked' appearance of that genus.

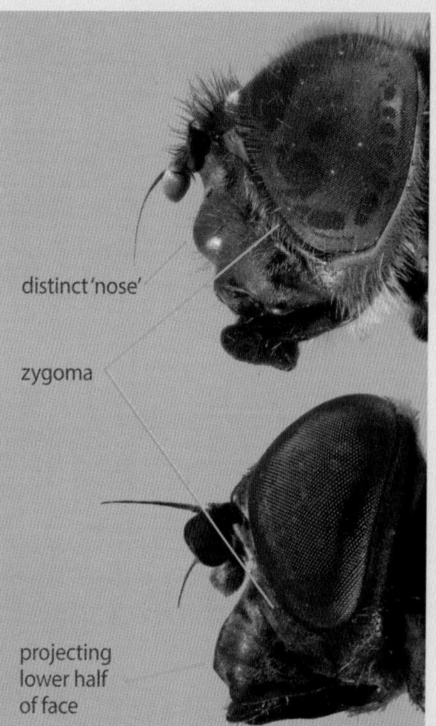

Cheilosia

38 British species (15 illustrated)

This genus has the largest number of species in Britain, all of which are basically 'black jobs' – although a few are furry. They are often dismissed by novices as too difficult to attempt but, in fact, many are actually quite distinctive and have a definite 'jizz'. A feature of the tribe Cheilosiini is the presence of the 'zygoma' – a sharply defined region of the face running along the eye margins. Since the other genera of Cheilosiini (*Ferdinandea*, *Portevinia* and *Rhingia*) have other distinctive features, the zygoma serves as a particularly good character for recognising *Cheilosia*. The larvae feed in the leaves, stems and roots of plants and a few are associated with fungi.

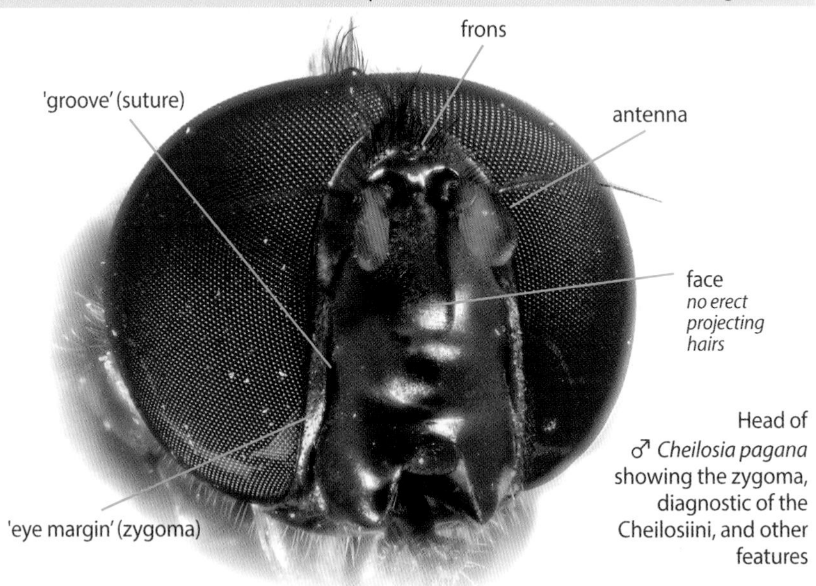

Head of ♂ *Cheilosia pagana* showing the zygoma, diagnostic of the Cheilosiini, and other features

Identifying Cheilosia

It does take some time to develop the knowledge and experience to recognise *Cheilosia* species in the field and reliable identification of the majority requires careful examination under magnification and good light.

Males and females often have different features, and some species have two broods that differ from one another. In addition females of some species have characters that are very similar to males of other species, so it is important to sex an individual, based on whether the eyes meet at the top of the head or not.

The specific characters used for identification are often rather small and subtle, and many are best evaluated in comparison to other species. Furthermore, it is often necessary to have accurately evaluated a suite of these subtle characters to reach an identification.

To add to the challenge, in some species these characters can be variable. Because of the variability and comparative nature of characters it is not always possible to assign an individual to a species.

The primary characters to assess are:

- whether or not the eyes are hairy (variable in some species)
- hairiness of the face
- leg colour (variable in some species)
- antenna colour

The table opposite divides up *Cheilosia* based on these primary features and indicates which species (***bold italic***) are covered by the species accounts. Full identification of all *Cheilosia* species is beyond the scope of this book. It requires detailed keys and access to voucher specimens for comparison.

More detailed information can be found in the species accounts.

Primary groupings of British *Cheilosia* species by feature

Wings with a dark cloud in the centre
page 162

C. illustrata ***C. caerulescens***

Front tarsi with segments 2,3,4 orange (to muddy brown!), contrasting with 1 and 5 which are black *page 174*

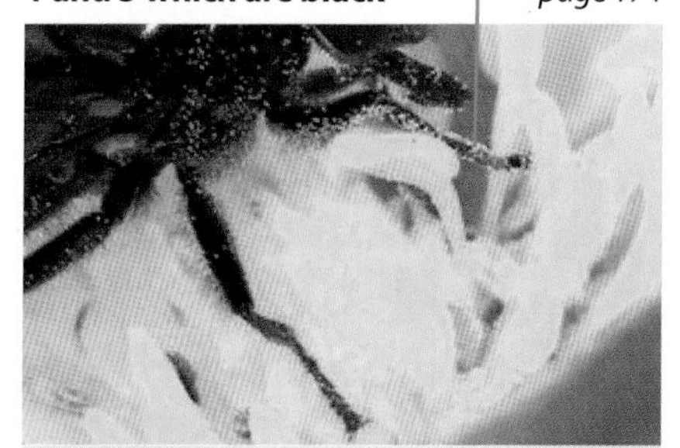

C. albitarsis ***C. ranunculi***
C. mutabilis [NS]

Face with erect, projecting hairs
page 164

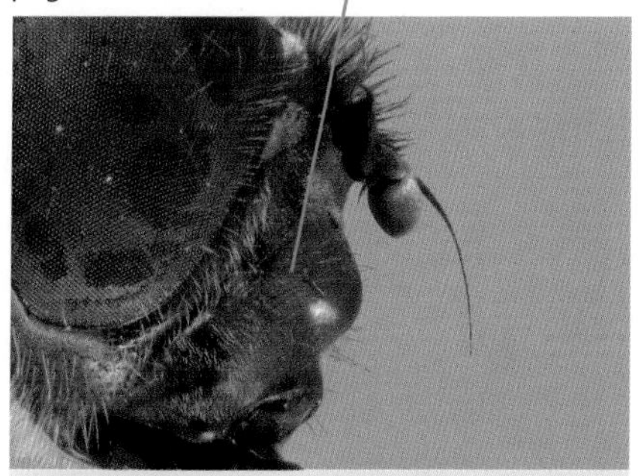

C. barbata [NS] *C. latifrons*
C. griseiventris ***C. variabilis***
C. lasiopa ***C. vulpina***

Eyes bare *page 166*

C. ahenea [VU] *C. pubera* [NS]
C. antiqua *C. sahlbergi* [VU]
C. longula ***C. scutellata***
C. nigripes [NS] *C. soror*
C. pagana *C. vicina*

Eyes hairy *pages 168–175*

C. albipila ***C. proxima***
C. bergenstammi *C. psilophthalma* [DD]
C. carbonaria [NS] *C. semifasciata* [NT]
C. chrysocoma *C. urbana*
C. cynocephala [NS] *C. uviformis*
C. fraterna *C. velutina* [NS]
C. grossa *C. vernalis*
C. impressa *C.* 'species B'
C. nebulosa

Cheilosia caerulescens

Wing length: 6·0–7·0 mm

Identification: A distinctive stout, black *Cheilosia* with a strong wing cloud, a distinctively projecting face without outstanding hairs (although heavily dusted) bare eyes and partially yellow legs. It is quite hairy on the thorax and abdomen. Adults tend to sit with folded wings, which can obscure the wing clouds.

Similar species: *Cheilosia illustrata*, but nowhere near as hairy and bee-like. The wing cloud is quite variable in extent and density and not always obvious when the wings are folded, so confusion with *C. proxima* (*p. 172*) and *C. vulpina* (*p. 164*) may be possible in photographs. The shape and dusting of the face, providing this is visible, should be conclusive in such cases.

Observation tips: The larvae mine the leaves of houseleeks, *Sempervivum* sp., causing obvious damage and it is perhaps easier to record these larval mines than the adults. There are two broods with adults flying in the spring and in mid-summer. They visit a variety of garden flowers. First recorded in Britain in 2006, almost certainly introduced in imported plants from the Continent. There are now numerous records across south-east England, from the Sussex coast to Bedfordshire. Experience from the Netherlands suggests it is likely to spread rapidly in urban gardens.

Cheilosia illustrata

Wing length: 8·5–10·25 mm

Identification: Although often described as a bumblebee mimic, it is not a very good one! At first glance, it might also be mistaken for *Leucozona lucorum* (*p. 112*) because of the band of long white hair across the base of the abdomen and the dark wing markings. However, the black scutellum and all black face should distinguish it (*L. lucorum* has a yellow scutellum). Amongst the *Cheilosia* only *C. illustrata* and *C. caerulescens* show a pronounced wing cloud, the latter differing in not being furry and having a protruding face.

Similar species: Aside from *Leucozona lucorum*, as it is, at least vaguely, a bumblebee mimic, it could possibly be confused with other mimics such as *Eriozona syrphoides* (*p. 112*) (especially when the black face of *C. illustrata* is not visible from a photo), *Eristalis intricaria* (*p. 206*) and *Volucella pellucens* (*p. 246*).

Observation tips: Strongly associated with Hogweed: larvae mine the stems and roots and adults are frequently seen visiting the flowers. However, other umbellifers are also visited and adults can therefore be encountered outside the flowering period of Hogweed. Common and widespread in lowland localities such as woodland edges and road verges, but also found in the uplands wherever Hogweed occurs.

See *Identifying wasp and bee mimics* on *pages 64–66*.

Cheilosia caerulescens × **6**

Cheilosia illustrata ♀ × **6**

Cheilosia variabilis

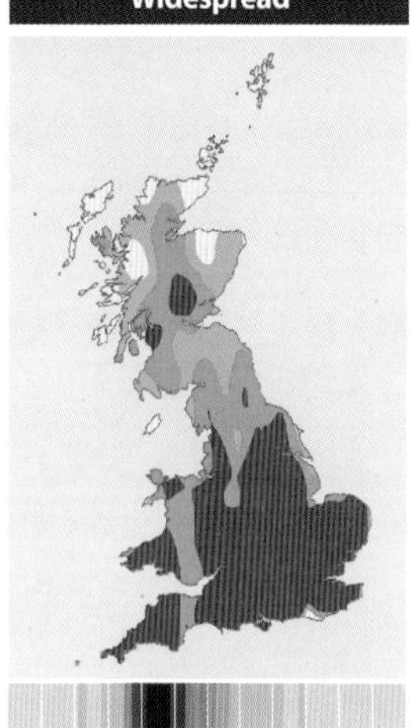

Wing length: 7·75–10·25 mm

Identification: A large and very black species with a long body and even longer wings. Adults often sit with their wings slightly parted giving it a delta-winged look. To be sure of the identification, it is important to check for distinct, upstanding hairs on the sides of the face below the 'nose'.

Similar species: Confusion with *Cheilosia lasiopa* (not illustrated), which is told by its shorter scutellar marginal bristles when viewed from the side, and *C. vulpina* (which is a similar shape but has partially yellow legs) is possible if using keys. In general shape and proportions, confusion is also possible with *C. griseiventris* (not illustrated), though this species has orange antennae.

Observation tips: Damp woodland, stream-sides and shady road verges throughout mainland Britain. Both sexes like to bask on sunny leaves, often on low-growing vegetation. The larvae feed in the stems and roots of Common Figwort.

Cheilosia vulpina

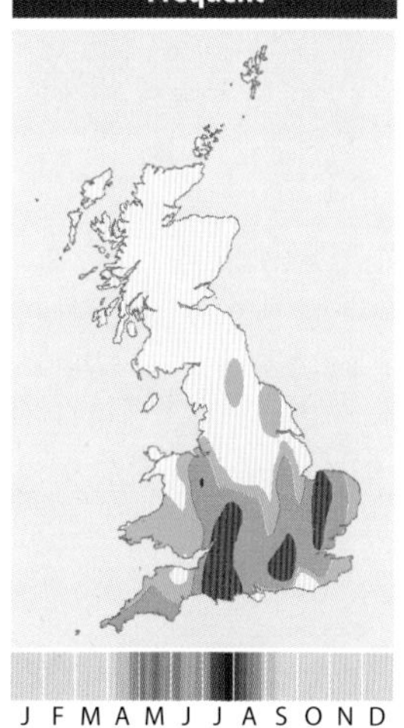

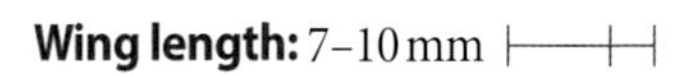

Wing length: 7–10 mm

Identification: A combination of distinct facial hairs below the 'nose', together with distinct yellow markings on the dark legs point towards this species. Females exhibit distinct hair stripes on the abdomen that are similar to those of female *C. proxima*.

Similar species: Confusion with species with dark antennae such as *C. variabilis*, *C. proxima* (*p. 172*) and *C. velutina* (not illustrated) is most likely, but also with other species with upstanding hairs on the face such as *C. lasiopa* and *C. griseiventris*.

Observation tips: This is mainly a woodland species which visits a wide range of umbellifers. It has two distinct generations: in May-June and in late July-August. This has been regarded as a scarce species, but seems to be becoming more common, especially in the East Midlands.

C. variabilis ♂
upstanding hairs on the side of the face
C. vulpina ♀
upstanding hairs on the side of the face
C. vulpina ♂
upstanding hairs on the side of the face
Cheilosia variabilis ♂ ×6
Cheilosia vulpina ♀ ×6
occiput heavily dusted
yellow markings on legs

Cheilosia pagana

Wing length: 4·75–8·5 mm

Identification: This is a distinctly shiny black hoverfly of variable size with orange antennae and hairless eyes. Males are slightly more difficult to separate from other *Cheilosia* than are females, although the orange antennae with black tips provide a good indication. The antennae of females tend to be disproportionately large making them rather more distinctive. Confusion is most likely with *C. bergenstammi* and *C. soror*, both of which have orange antennae.

Similar species: *C. soror* (not illustrated), *C. caerulescens* (*p. 162*); and *C. bergenstammi* and *C. fraterna* (both *p. 170*) – both of which have hairy eyes.

Observation tips: A common and widespread, multi-brooded species which occurs throughout the season. It can be found at Lesser Celandine early in the Spring, at Cow Parsley in early Summer and at Ivy in the Autumn. Larvae have been found amongst the rotting roots of Cow Parsley.

Cheilosia scutellata

Wing length: 6–9 mm

Identification: This is a moderately large and somewhat elongated *Cheilosia* that has bare eyes, dark antennae and a very distinct bulging 'nose'. It should not realistically be confused with any others apart from *C. soror* because the antennae of *C. scutellata* can occasionally be quite pale. In general shape and 'jizz', the most likely confusion is with *C. soror* and possibly large *C. pagana*.

Similar species: *C. longula* (see *opposite*), but in general shape and 'jizz', the most likely confusion is with *Cheilosia soror* (not illustrated) and possibly large *C. pagana*, both of which have orange antennae. However, confusion is also possible with *C. barbata* (not illustrated) and perhaps even *C. bergenstammi* (*p. 170*) if the antennal colours are not visible in photographs.

Observation tips: Mainly a woodland species which visits umbellifers in mid-Summer. The larvae feed in *Boletus* fungi – a habit that is shared with the similar *C. longula*. Large numbers of larvae (up to 50) can be found in a single *Boletus* cap and their feeding causes the fungus to liquefy, resulting in a patch of brown slime being all that is left.

Cheilosia pagana ♀ ×**6**

Cheilosia scutellata ♂ ×**6**

Cheilosia albipila

Frequent

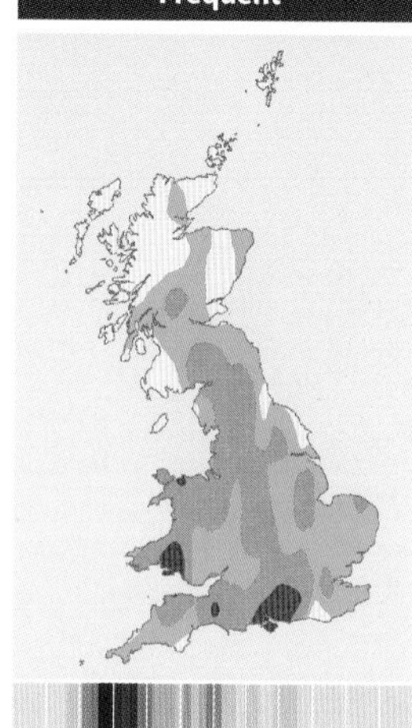

J F M A M J J A S O N D

Wing length: 8·75–10·75 mm

Identification: The combination of entirely orange hind tibiae, orange antennae and orange fur is reasonably distinctive.

Similar species: There are several other species with which it might be mistaken, the most likely being *Cheilosia nebulosa* (not illustrated), which is generally smaller, is slightly less furry and has more pronounced wing shading. Other possible confusion species include *C. grossa*, which is a similar size and has orange hind tibiae, but is more furry and has black antennae and *C. fraterna* (*p. 170*), in which the antennae and hind tibiae are also orange but is not furry

Observation tips: Because it flies so early in the Spring, this species is often overlooked. Males visit willows and other flowering shrubs, whilst females can be found egg-laying around Marsh Thistle rosettes. The larvae feed in the stems of thistles and, like *C. grossa*, it is more easily recorded, especially in upland areas, by searching for larvae in thistle stems during July (see *page 23*).

Cheilosia grossa 

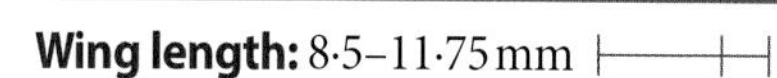

Frequent

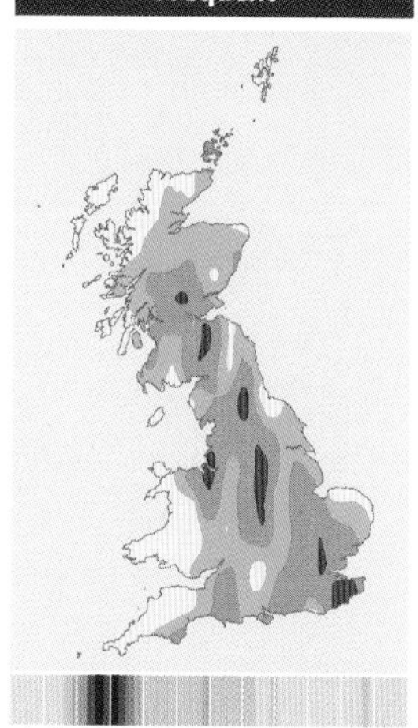

J F M A M J J A S O N D

Wing length: 8·5–11·75 mm

Identification: A large, furry hoverfly with orange tibiae and black antennae. This combination of characters makes it hard to confuse with any other *Cheilosia*.

Similar species: Although *C. grossa* is unlikely to be confused with any other species, specimens with a strong black ring on the hind tibia have been noted. Such problem specimens may key out to the PROXIMA group in Stubbs & Falk and to the CANICULARIS group in van Veen. However, the absence of bristles on the scutellum helps to confirm that a specimen falls into the GROSSA group

Observation tips: An early Spring species which can be found visiting willows and other flowering shrubs. Males are noted for hovering at or above head-height in open ground. The larvae feed in the roots and stems of Marsh and Spear Thistle where their presence is betrayed by the characteristic multi-stemmed form of the affected plant. Because of its early flight period, adults are often overlooked and it is more readily found by searching for larvae in thistle stems during July and early August (see *page 23*). Larval records show that it is much commoner and more widespread than records of adults suggest.

Female *C. albipila* egg-laying on Marsh Thistle rosette – see *page 23*.

Cheilosia albipila ♀ ×**6**

♀ – *Cheilosia grossa* ×**6** – ♂

Cheilosia bergenstammi

Widespread

Wing length: 7·25–9·25 mm

Identification: A relatively large *Cheilosia* with orange antennae and eyes that are usually densely hairy. Care needs to be taken, however, because the level of eye hairiness is variable and it may be confused with several other species that have bare eyes and orange antennae, such as *C. soror.* The tarsi are extensively pale and the tibiae are usually orange with black rings, although some individuals can have paler legs like *C. fraterna.*

Observation tips: There are two generations, one in the Spring and one in high Summer. The Spring generation is often found at Dandelion flowers, whilst the Summer brood is commonly found visiting ragworts, the stems and roots of which are mined by the larvae. This is a very widespread species which occurs throughout the British Isles.

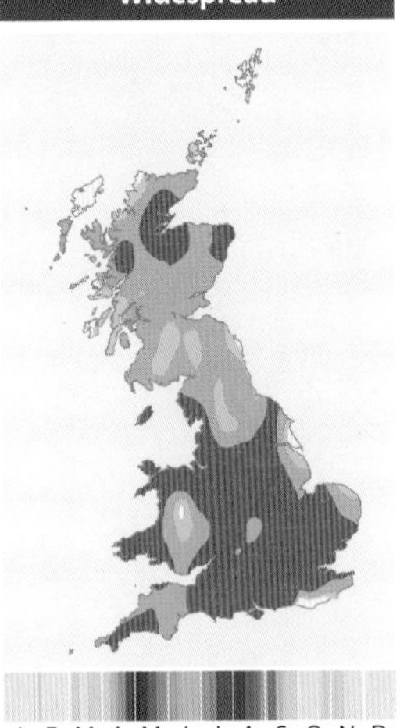

Similar species: Some individuals of the very similar *C. fraterna* and *C. bergenstammi* can show variations in leg colour which make this feature difficult to rely on for identification. In these cases other characteristics (see *C. fraterna* below) need to be evaluated. These two species are themselves similar to a number of other species – see table on *page 161*.
Other species such as *C. albipila* (*p. 168*), *C. nebulosa* (not illustrated) and *C. grossa* (*p. 168*) may be confused at first glance – although the furriness of these species should be distinctive. *C. pagana*, *C. scutellata* (both *p. 166*) and *C. soror* (not illustrated) may also cause confusion because they have orange antennae, though the eyes of these species are not hairy (but beware *C. bergenstammi* can have almost hairless eyes). Confusion with *C. albitarsis* (*p. 174*) and *C. carbonaria*, *C.antiqua* and *C. vicina* (none illustrated) is also possible in photographs which do not clearly show the dark antennae and leg characters of these species.

Cheilosia fraterna

Widespread

Wing length: 6·25–10·25 mm

Identification: Males and females show strong sexual dimorphism although both sexes have orange antennae and a wholly orange hind tibia. Males have a distinct area of black hairs towards the middle and rear of the thorax; females have very short and closely adpressed golden hairs in this area.
The most likely species with which *C. fraterna* can be confused is *C. bergenstammi*, which normally has a dark ring on the tibiae, but also has much longer scutellar hairs in the male (plus more extensive long hairs on the underside of the hind femora) and much longer thoracic hairs in the female.

Observation tips: Most records are from wet grasslands and upland situations. This species is more frequent in the north and west where it can be quite common.

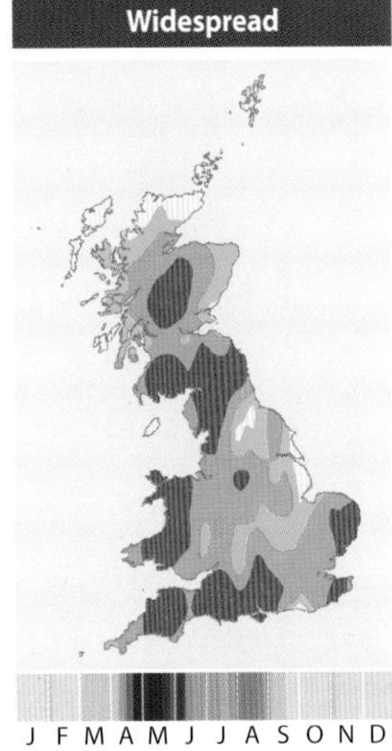

Cheilosia bergenstammi ♂ ×**6**

Cheilosia fraterna ♀ ×**6**

Cheilosia impressa

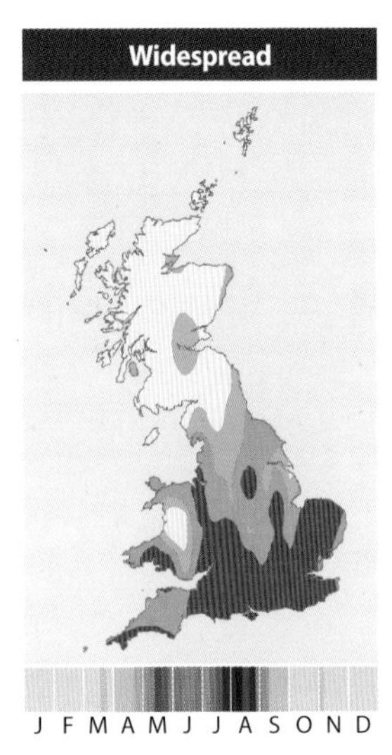

Wing length: 5·75–8 mm

Identification: Females are reasonably distinctive because they are one of the few *Cheilosia* with distinct yellow bases to the wings. Males are very dark blue-black with obviously red eyes and could therefore be confused with *Chrysogaster solstitialis.* The yellow wing bases of females could lead to confusion with *Chrysogaster cemiteriorum* and perhaps even *Myolepta dubia* whose yellow markings show through the wings in a similar manner.

Similar species: Unless looked at closely *Chrysogaster cemiteriorum* and *C. solstitialis* (both *p. 188*) and *Myolepta dubia* (*p. 197*) appear similar, but all these species lack a zygoma.

Observation tips: Mainly associated with woodland edges, rides and sometimes roadside verges during the peak umbellifer season (late May through to August). It is primarily southern in distribution but occurs north to the Scottish borders.

Cheilosia proxima

Wing length: 6·25–8·5 mm

Identification: One of a suite of *Cheilosia* species that are very similar. This is a polymorphic species that has two distinct forms: Spring and Summer. The combination of hairy eyes, partially pale legs and dusted underside to the abdomen points towards this species. In theory it should only be confused with *C. velutina*, but is not straightforward to identify.

Similar species: *Cheilosia velutina* (not illustrated) and *C. vulpina* (*p. 164*) are the most likely sources of confusion. As *C. velutina* is almost identical to *C. proxima*, this species can never be ruled out. *C. vulpina* is readily separated if the facial hairs can be seen but can be impossible to separate from photographs which do not show this feature.

Observation tips: This is a regular visitor to umbellifer flowers and occurs widely in rough grassland with thistles and along woodland rides; the larvae are stem miners in Creeping Thistle. It is widespread across mainland Britain.

♀ – *Cheilosia impressa* × **6** – ♂

♀ – *Cheilosia proxima* × **6** – ♂

the dusted underside of the abdomen is indicative of *C. proxima*

Cheilosia chrysocoma

Nationally Scarce

Local

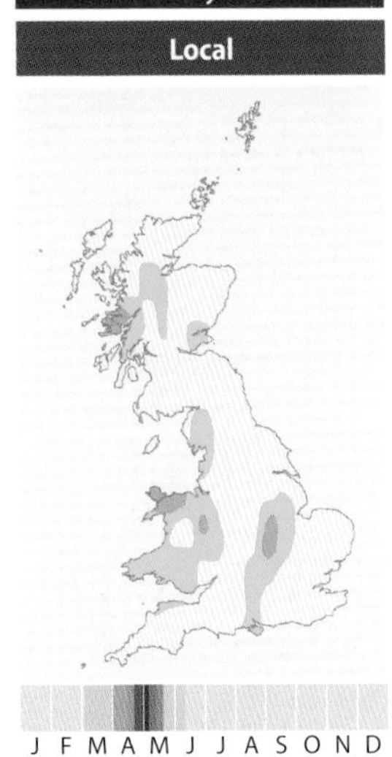

Wing length: 8–10·25 mm

Identification: A highly distinctive mimic of the female of the solitary mining bee *Andrena fulva* (inset). The body hairs are long and red-orange and the underlying body colour is similar, making it difficult to confuse with any other hoverfly. Consequently, this is one of the few *Cheilosia* that can be readily identified from a photograph.

Similar species: Some colour forms of *Merodon equestris* (*pp. 31* and *222*) are very similar. They can be distinguished by the loop in wing vein R_{4+5} and the presence of a large tooth on the enlarged hind femur, which are absent in *C. chrysocoma*.

Observation tips: A rare species with few, scattered records, often from woodland edges. Early in the year, and in such locations, adults basking on sunny leaves look remarkably like *A. fulva*. The female has been seen ovipositing on Angelica, which is believed to be the larval food plant.

CHEILOSIA WITH ORANGE AND BLACK TARSAL SEGMENTS

Cheilosia albitarsis / ranunculi

Widespread / Frequent

Wing length: 7·0–9·5 mm

Identification: Males have hairy eyes, but females do not – however both sexes of these two species are distinctive as they have dark antennae and entirely black legs, except for the first two or three tarsal segments on the front and middle legs, which are yellow (although they can be a murky orange-brown).

Similar species: Apart from this species pair the Nationally Scarce *C. mutabilis* (not illustrated) is very similar, differing in its dark halteres. Also *C. bergenstammi* and *C. fraterna* (both *p. 170*), which both have orange antennae and generally paler legs; *C. carbonaria*, *C. semifasciata* and *C. vicina*, and females of bare-eyed species such as *C. antiqua* and *C. nigripes* (none illustrated).

Observation tips: Strongly associated with buttercups; the larvae feed around the roots and adults visit the flowers. *C. albitarsis* occurs throughout Britain, whilst *C. ranunculi* is scarce in northern England and there are very few Scottish records (the map shows their combined distributions). Both occur in damp fields with buttercups and they often occur together at the same site and time. The ecological separation between them has yet to be resolved.

Tawny Mining-bee *Andrena fulva*, which *Cheilosia chrysocoma* mimics

Cheilosia chrysocoma ♂ ×**6**

Cheilosia albitarsis ♂ ×**6**

paler tarsal segments 2–4 contrast with black segment 5

C. ranunculi ♂ front tarsus segment 5 tapers towards the claws

C. albitarsis ♂ front tarsus segment 5 parallel-sided

The most reliable way of separating *C. albitarsis* from *C. ranunculi* is by examination of the male genitalia, although females cannot be separated in this way.

Ferdinandea

2 British species (1 illustrated)

An easily recognised genus in which the thorax has a pair of broad, grey, longitudinal stripes and the abdomen has a somewhat metallic sheen. There are strong black bristles on the sides of the thorax, which is an unusual feature amongst hoverflies. The larvae live in sap runs.

Ferdinandea cuprea

Widespread

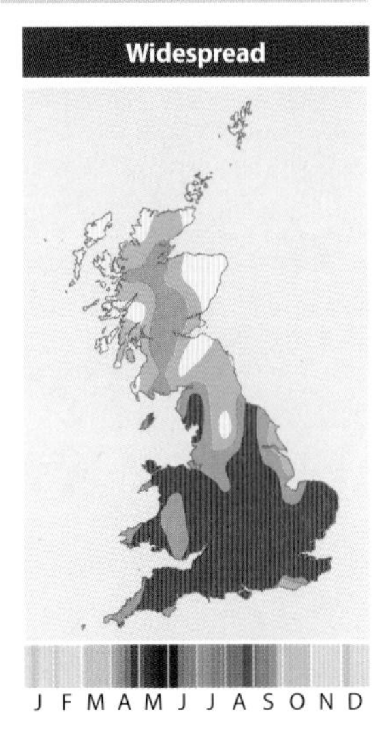

Wing length: 7·5–11·25 mm

Identification: An instantly recognisable and very attractive hoverfly with a metallic, brassy abdomen, grey stripes running along the thorax, wing markings and yellow legs. It is unusual for a hoverfly because it bears a series of strong bristles on the sides of the thorax.

Similar species: Confusion is most likely with the Nationally Scarce *Ferdinandea ruficornis* (not illustrated), which lacks strong spines on the tibiae and has red, rather than black, aristae.

Observation tips: A woodland species which is frequently found basking on sunlit leaves or on the trunks of trees (and even wooden posts and telegraph poles); they are less often seen visiting flowers. Widespread, especially in the south, but rarely abundant.

Portevinia

1 British species (illustrated)

A dark species with silver-grey abdominal markings. The larvae mine the bulbs and stem bases of Ramsons. Adult emergence is timed to coincide with the flowering of Ramsons and they are rarely found far from the plant. They visit the flowers and bask on the sunlit foliage where males tend to sit with their wings held in a very characteristic delta shape.

Portevinia maculata

Frequent

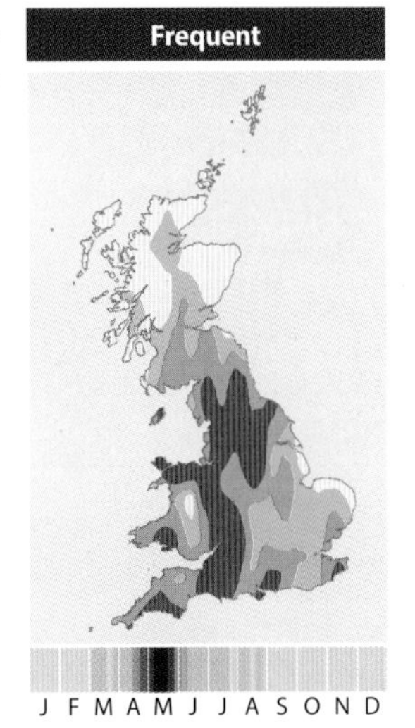

Wing length: 6–8·25 mm

Identification: The combination of orange antennae, blunt facial profile and rather square, grey markings on the abdomen – together with the close association with Ramsons – make this species relatively easy to identify.

Similar species: *Cheilosia* (*pp. 160-175*) with orange antennae. *C. fasciata* is a very similar European species which also breeds in Ramsons, but it has not been found in Britain. Experience has shown that *Platycheirus albimanus* (*p. 80*) can be confused with this species.

Observation tips: Confined to locations where patches of Ramsons grow (*e.g.* woodlands, mature hedgerows and large gardens). Where Ramsons is abundant, this hoverfly is usually also abundant. Most records are of males; as females tend to stay low down amongst the plants and are not so readily seen.

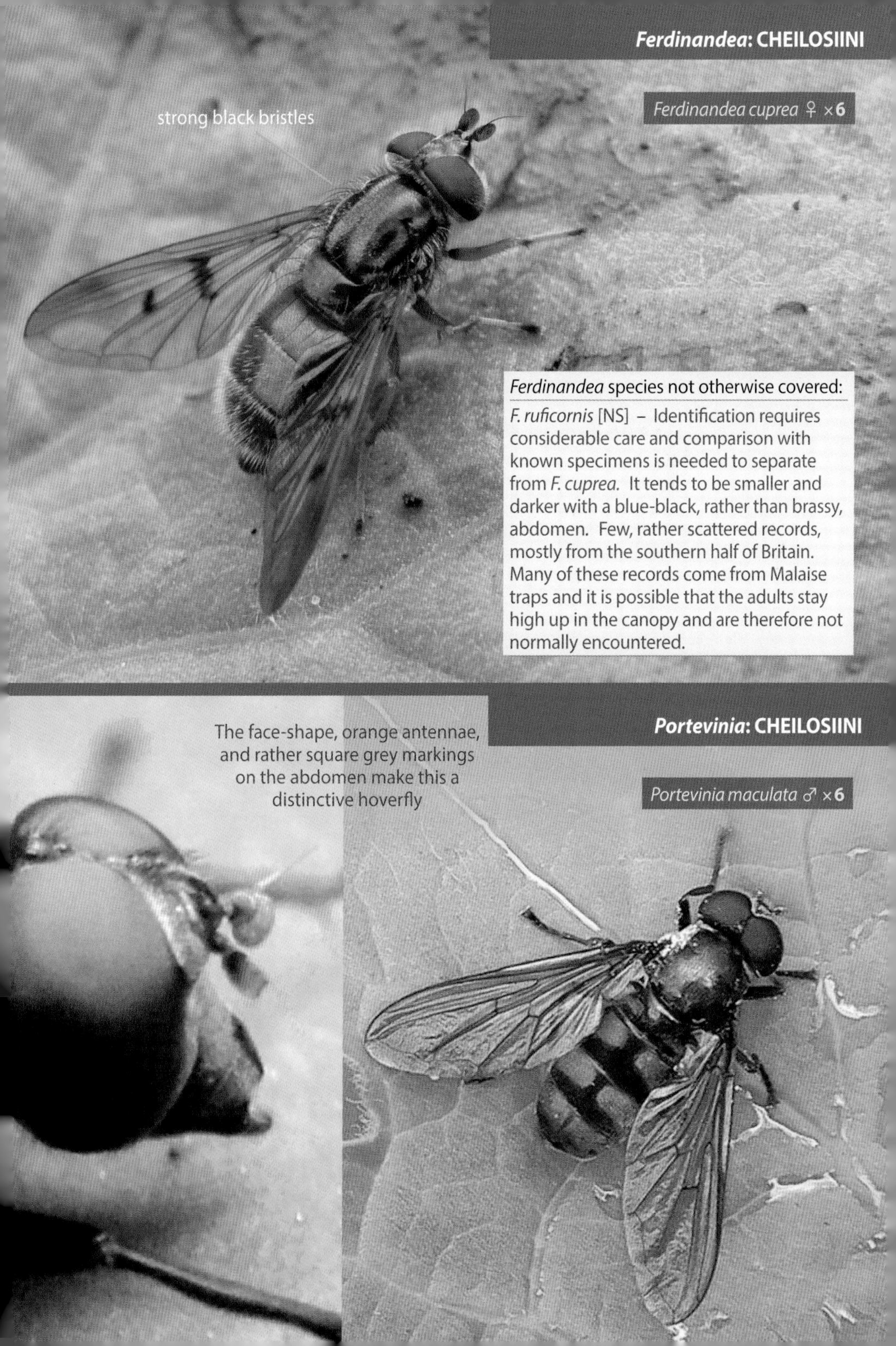

Ferdinandea species not otherwise covered:

F. ruficornis [NS] – Identification requires considerable care and comparison with known specimens is needed to separate from *F. cuprea.* It tends to be smaller and darker with a blue-black, rather than brassy, abdomen. Few, rather scattered records, mostly from the southern half of Britain. Many of these records come from Malaise traps and it is possible that the adults stay high up in the canopy and are therefore not normally encountered.

Portevinia: **CHEILOSIINI**

Rhingia

2 British species (both illustrated)

An unmistakable genus of dumpy orange hoverflies with a very obvious, exceptionally long rostrum that encloses the proboscis and allows the fly to feed on nectar and pollen in deep flowers, such as Red Campion, which other hoverflies cannot reach. The larvae of *Rhingia campestris* live in fresh cattle dung, but have also been found in other highly enriched wet media such as silage. Its abundance in areas with few cattle, such as East Anglia, suggests that breeding habitats other than cow dung must be significant.

Rhingia campestris

Wing length: 6–9·5 mm

Identification: The only species with which it can be confused is *R. rostrata*. In *R. campestris*, the edges of the abdomen are darkened and the legs, face and scutellum are all duller and darker in colour.

Similar species: Only *Rhingia rostrata*.

Observation tips: A common and widespread species with two generations, one flying in May–June and the other in mid- to late Summer. It is most common in woodland and field edges but can be found in almost all habitats. It visits a wide range of flowers, including those with deep tubes, such as Red Campion and Bluebell, which other hoverflies cannot utilise.

Rhingia rostrata

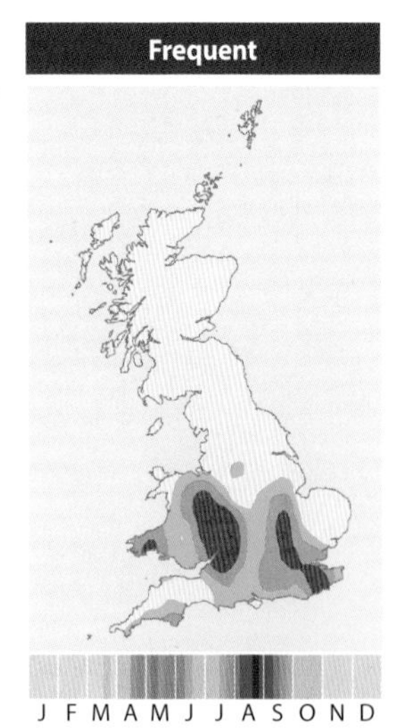

Wing length: 7·5–9·5 mm

Identification: Whilst this species can be overlooked amongst the much commoner *R. campestris* (with which it occurs), it actually looks distinctly different. The thorax is somewhat more bluish, there are no dark edges to the abdomen, and it is altogether more brightly coloured with the face, legs and scutellum orange-yellow rather than dull brown.

Similar species: Only *Rhingia campestris*.

Observation tips: Formerly a highly localised species that was confined to the Weald, the Welsh borders and South Wales. It has, however, undergone a major range expansion and is now found more widely across southern Britain to North Wales and also in the East Midlands. It seems to be more strongly confined to woodland (especially semi-natural woodland) than *R. campestris*, but since little is known about its larval biology the reasons are not understood.

Rhingia campestris ♀ ×**10**

Rhingia rostrata ♀ ×**10**

The two *Rhingia* species can be separated as follows:

***R. campestris*:** generally darker with longer rostrum than *R. rostrata*; ABDOMEN: edge **darkened**

***R. rostrata*:** generally paler with shorter rostrum than *R. campestris*; ABDOMEN: edge **not darkened**

Guide to Chrysogastrini

This is a somewhat heterogeneous grouping of 10 genera of small to medium-sized, dark hoverflies. Several genera are distinctive but difficult to find. These include *Brachyopa*, *Myolepta*, *Neoascia* and *Sphegina*. The hairy humeri are not covered by the head and are usually easy to see as the thorax hairs are relatively short by comparison. In several genera however these hairs may be small and sparse, and can be bristle-like; as such they are easily overlooked. **The most useful feature for distinguishing this tribe is the shape of the face; unlike many other dark hoverflies it is often strongly concave and does not have a nose-like 'knob'.** However, there are exceptions, especially within *Chrysogaster* and *Melanogaster*. Several genera have quite a strong metallic sheen, especially *Lejogaster*, *Orthonevra* and *Riponnensia*.

1

a) Abdomen distinctly 'waisted'

See also couplet 1 in Guide to Bacchini (*p. 72*)

Neoascia p. 182

Sphegina p. 184

b) Abdomen 'normal'

Sphegina clunipes

2

from **1b**

a) Abdomen distinctly metallic

Beware - the soldierfly *Chloromyia formosa* may be mistaken for *Lejogaster.* Check the wing venation to make sure it is a hoverfly.

Lejogaster p. 192

b) Abdomen with metallic or shining margin distinct from dull upper surface

c) Abdomen non-metallic

3 from **2b**

a) Thorax metallic bronzy
Abdomen segment edges: T1–T4 shiny

See accounts for differentiating features.

Orthonevra *p. 190*

Riponnensia *p. 190*

b) Thorax dark, often matt or with greenish/purplish reflections
Abdomen segment edges:
T1 dull; T2–T4 shiny

▶ 4

4 from **3b**

a) Antennae black

Melanogaster *p. 186*

b) Antennae orange to brown

Beware - some females can have rather darkened antennae.

Chrysogaster *p. 188*

5 from **2c**

a) Abdomen with yellow markings

Myolepta *p. 197*

Beware - the yellow markings may not always be obvious, especially when the wings are closed over the back.

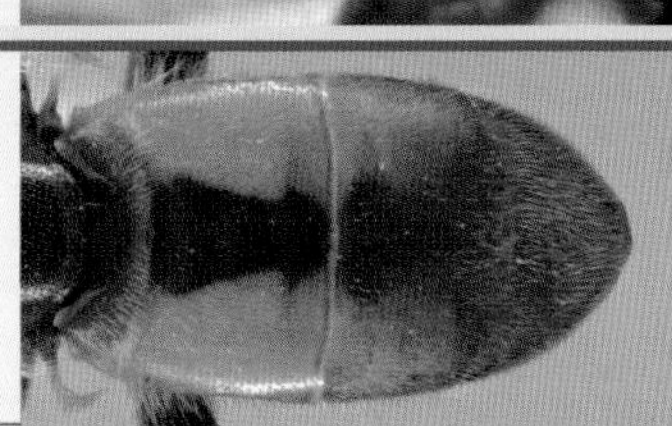

Myolepta dubia

b) Abdomen orange-brown
Thorax orange-brown

The abdomen may darken to deep red-brown with age.

Hammerschmidtia *p. 196*

c) Abdomen orange-brown
Thorax grey

Brachyopa *p. 194*

Brachyopa insensilis

Neoascia

6 British species (3 illustrated)

This genus of tiny, black hoverflies has narrow, wasp-like waists, especially in the females. Males are less obviously waisted but are somewhat elongate and often have pronounced genital capsules. Larvae have been found in wet manure and compost and accumulations of decaying vegetation around ponds and ditches.

Similar species: Outside of *Neoascia*, other species that may possibly cause confusion are *Baccha elongata* (*p. 74*), the hind femora of which are not enlarged, and *Sphegina* (*p. 184*) which are larger, more elongate and have hind femora that are more expanded.

Neoascia meticulosa / tenur

Frequent / Widespread

J F M A M J J A S O N D

Wing length: 3–5·5 mm

Identification: The outer cross-veins of the wing are clear and abdominal markings are often lacking or indistinct. These two species are separated by the presence (*N. meticulosa*) or absence (*N. tenur*) of a yellow tip to the hind femur, but this can be indistinct. A third species, *N. geniculata*, is often confused with *N. tenur* as the differences between the species are very subtle and are based on the shape of antenna segment 3 in males and the more waisted shape of abdomen segment T2 in females. It may not be possible to identify some individuals to species.

Similar species: *Neoascia geniculata* (not illustrated).

Observation tips: Wet places such as ditch and pond margins where they can occur in considerable numbers. Adults visit low-growing flowers such as Tormentil and Creeping Buttercup. The two species often occur together and it is not clear whether they have different habitat preferences. *N. tenur* perhaps occurs more frequently in acidic wetlands. The map shows their combined distribution.

Neoascia podagrica

Widespread

Wing length: 3·5–5 mm

Identification: The outer cross-veins of the wing are clouded and abdomen segment T2 has distinct yellow markings. The species with which it is most likely to be confused is *N. obliqua*, which is much scarcer and has oblique markings on abdomen segment T2. *N. obliqua* tends to be found around beds of Butterbur, although both species can occur together.

Similar species: *N. obliqua* (see above) and the Nationally Scarce *N. interrupta* (not illustrated), which is unique within the genus in having spots on the side of abdomen segment T4, though these markings are sometimes obscure.

Observation tips: A widespread and common species in a range of habitats such as woodland rides, hedgerows and around ponds and ditches. It is often found visiting low-growing flowers and has a long flight season.

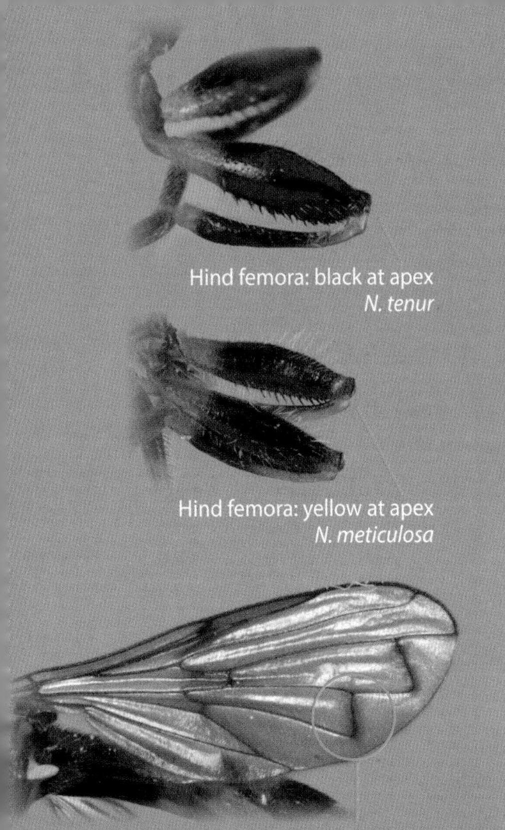

Hind femora: black at apex
N. tenur

Hind femora: yellow at apex
N. meticulosa

Wings: cross-veins clouded
N. podagrica

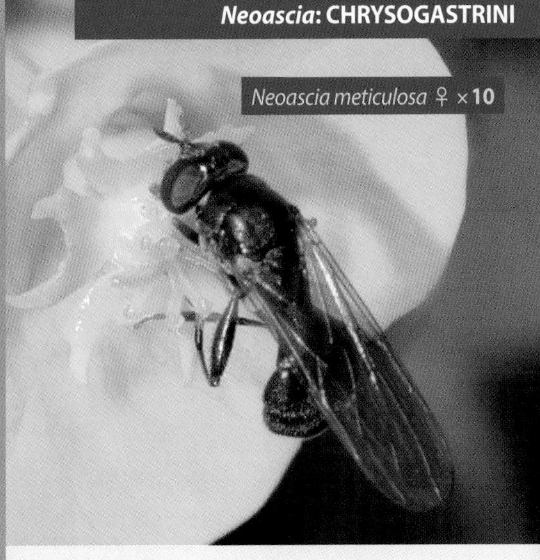

Neoascia meticulosa ♀ ×**10**

Neoascia species not otherwise covered:

N. geniculata – Widespread, but scarce. Possibly under-recorded because identification is tricky, relying on rather subtle differences in antennal dimensions.

N. interrupta [NS] – Relatively recently discovered in Britain (1981) and, so far, recorded mainly from the east of England.

N. obliqua – A widely distributed but scarce species. It was thought to be rare, but since it was discovered to be associated with beds of Butterbur, it has been found more frequently.

♂ – *Neoascia podagrica* ×**10** – ♀

Sphegina

4 British species (3 illustrated)

Small, delicate, 'wasp-waisted' hoverflies that tend to fly with their legs dangling downwards. Most frequently encountered in dappled light around woodland flowers such as Angelica, Pignut or Sanicle. They are unlikely to be confused with any other genera apart from *Neoascia* (*p. 182*) or *Baccha* (*p. 74*). They tend to be more abundant in the north and west, probably because they are particularly associated with damp habitats. The larvae live in decaying sap under bark, usually in damp situations such as logs lying on wet ground or partly submerged in pools or streams.

Similar species: The four *Sphegina* species are very similar. *Baccha elongata* (*p. 74*), the hind femora of which are not enlarged, and *Neoascia* species (*p. 182*), which are smaller, less elongate and have hind femora that are less expanded, may also possibly cause confusion.

Sphegina clunipes

Widespread

Wing length: 4·75–7 mm

Identification: There are three very similar species: *S. clunipes*, *S. elegans* and *S. verecunda*. *S. elegans* has yellow rather than dark humeri, and male *S. clunipes* has a clearly visible pointed process on the genitalia, which *S. verucunda* lacks. Separating females of *S. clunipes* and *S. verecunda* is difficult and requires careful examination of the wing venation.

Observation tips: The commonest member of the genus and can be numerous, especially in Scotland. Found around and underneath white umbellifers in damp places in woodlands and hedgerows. In some northern coniferous woodlands it is also frequent at Tormentil flowers.

Sphegina sibirica

Frequent

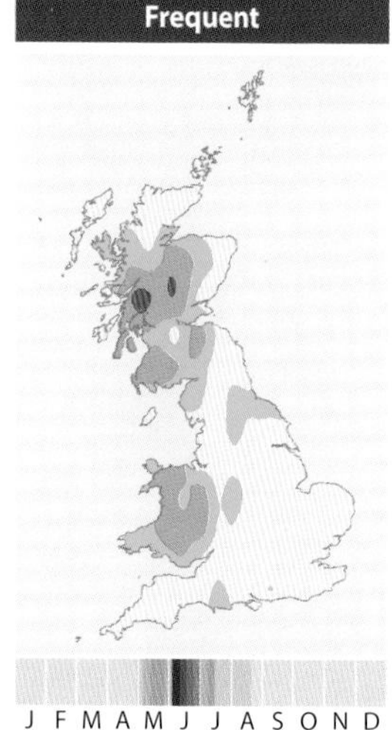

Wing length: 7 mm

Identification: Averages slightly larger than *S. clunipes*, with shining patches on the sides of the thorax and black 'feet', contrasting with the pale tibia. The underside of the base of the abdomen is entirely membranous, with the 1st sternite absent or reduced. Very variable in colour from all dark to all yellow, with some having yellow-and-black piebald patterns. Whilst completely pale forms of *S. elegans* do occur, piebald colour forms appear to be unique to *S. sibirica*.

Observation tips: A northern and western species which is most frequently found in coniferous woodlands. It is a relatively recent colonist that is rapidly extending its range and is likely to become more widespread. Its powers of dispersal are quite remarkable and it can be found on offshore islands and well above the tree-line on Scottish mountains.

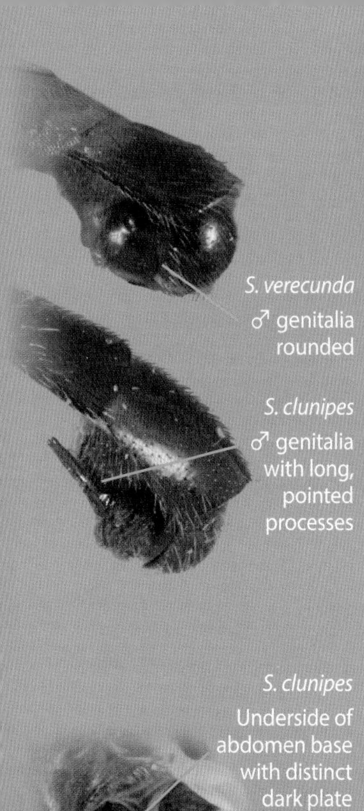

Sphegina clunipes ♀ ×**8**

S. clunipes
Underside of abdomen base with distinct dark plate

Sphegina species not otherwise covered:

S. elegans – Like *S. clunipes*, but with yellowish humeri (see *inset above*) – beware some *S. clunipes* have slight yellow marks on the humeri – and shorter processes on the male genitalia. Widespread and not uncommon but less frequent than *S. clunipes* especially in Scotland where there are very few records.

S. verecunda – Males are easily distinguished from *S. clunipes* by the genital character (see *above left*). Females are more difficult to separate. The scarcest *Sphegina* and generally found in wetter areas of woodland.

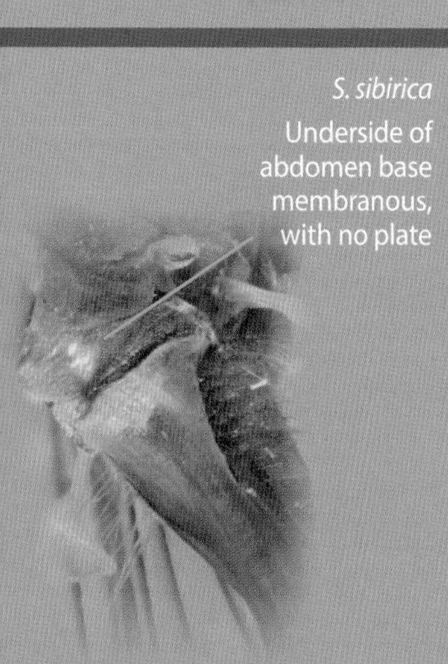

S. sibirica
Underside of abdomen base membranous, with no plate

Sphegina sibirica ♀ ×**8**

Melanogaster

2 British species (1 illustrated)

These are small black hoverflies in which the female does not have a 'nose' and that of the male is poorly developed. The abdomen is rather flattened with a matt centre contrasting with shiny edges (most noticeable in females). Identification can be difficult, requiring good magnification and lighting to separate the two species. They occur in wet meadows and along flowery heathland edge. The larvae live among emergent plants at the edges of ponds and ditches.

Melanogaster hirtella

Wing length: 5–6 mm

Identification: Females are more straightforward to identify than males. Their facial profile has no indication of a 'nose', their antennae are completely black; their abdomen is matt black with shiny edges (segments T2–T4 only) and there are upright yellowish hairs on the thorax. Males are more difficult to identify because they do have a slight 'nose' and are easily mistaken for small *Cheilosia*, particularly *C. antiqua*.

Similar species: *Melanogaster aerosa* (not illustrated) is very difficult to separate and requires comparative specimens. Male *Cheilosia antiqua* (not illustrated), is also similar in having completely black legs and hairless eyes but can be differentiated by the presence of a zygoma.

Observation tips: A widespread and abundant Spring species, generally occurring in damp meadows where it visits buttercups and other flowers.

Differentiating features of the partially metallic and metallic Chrysogastrini (*see opposite*)

GENUS	THORAX and ABDOMEN	OTHER FEATURES
Melanogaster *above*	THORAX: dark, often matt or with greenish/purplish reflections ABDOMEN: Flattened with a matt centre contrasting with shiny edges to **T2–T4** with **T1 dull**	ANTENNA: **black**
Chrysogaster *p. 188*		ANTENNA: **brown/orange**, though some may be darker
Orthonevra *p. 190*	THORAX: bronzy metallic ABDOMEN: Flattened with a matt centre contrasting with shiny edges to **T1–T4**	FACE: dust bands beneath antennae **thin** – see *page 191*
Riponnensia *p. 190*		FACE: dust bands beneath antennae **broad** – see *page 191*
Lejogaster *p. 192*	ABDOMEN: **wholly metallic**	

See *opposite* for comparisons of *Melanogaster/Chrysogaster*, *Lejogaster* and *Orthonevra/Riponnensia* abdomens.

M. hirtella males have a slight 'nose'

Melanogaster hirtella ♀ ▲ ♂ ▼ ×**10**

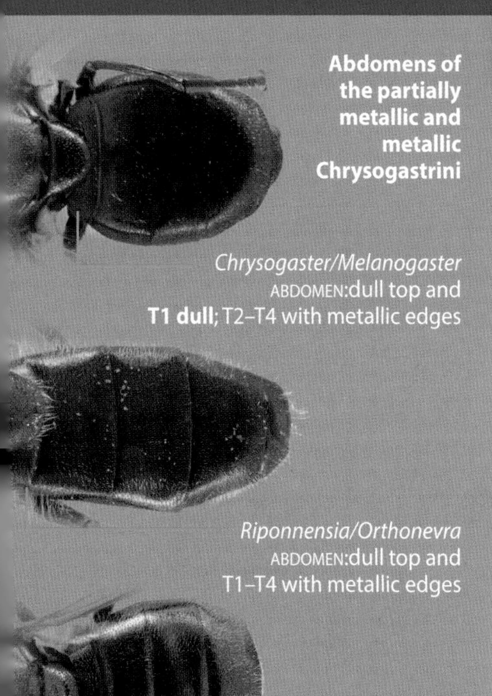

Abdomens of the partially metallic and metallic Chrysogastrini

Chrysogaster/Melanogaster
ABDOMEN: dull top and **T1 dull**; T2–T4 with metallic edges

Riponnensia/Orthonevra
ABDOMEN: dull top and T1–T4 with metallic edges

Lejogaster
ABDOMEN: wholly metallic

Melanogaster species not otherwise covered:

M. aerosa – Mainly northern and western in acidic wetlands, but rarely abundant. The separation from *M. hirtella* is tricky, involving subtle differences in hair colours.

Chrysogaster

3 British species (2 illustrated)

Small, shiny black hoverflies with somewhat metallic reflections. Closely related to *Lejogaster* (*p. 192*), *Melanogaster* (*p. 186*), *Orthonevra* and *Riponnensia* (both *p. 190*), which are all broadly similar in general appearance. However, confusion is most likely with *Cheilosia* (*pp. 162–175*), which are not metallic and also generally have a much more pronounced 'nose' in the centre of the face. They are flower visitors and can occur in considerable numbers. The larvae are aquatic and live in mud and among decaying vegetation at the edges of ponds and ditches.

Chrysogaster cemiteriorum

Wing length: 5·25–6·5 mm

Identification: The combination of slightly metallic abdomen and body and strongly yellow wing bases, makes this a moderately straightforward species to detect. The sides of the thorax above the front coxae are usually grey-dusted.

Similar species: Within this genus, it is most likely to be confused with *C. virescens* (not illustrated), which has a more metallic green hue but is best separated by the lack of dusting on the sides of the thorax, but this is variable in *C. cemiteriorum* so care is needed. Confusion at rest is possible with *Cheilosia impressa* (*p. 172*) and *Myolepta dubia* (*p. 197*); both have, or give the impression of, strongly yellowed wing bases (in *M. dubia* this comes from yellow markings on abdomen segments T2 and T3 that are visible through the wings).

Observation tips: This is a woodland species that visits a range of umbellifers in mid-Summer. It generally occurs in much lower numbers than *C. solstitialis* with which it is often found.

Chrysogaster solstitialis

Wing length: 6–7·25 mm

Identification: A small and very dark looking hoverfly with a matt black body with purplish reflections on the thorax (depending on the light angle) and strongly darkened wings. Males have strikingly bright-red eyes giving them a very distinctive appearance.

Similar species: Other *Chrysogaster* (see above); confusion is also possible with some housefly-like species in other families, which are rather black and with bright-red eyes.

Observation tips: A common species of woodlands, road verges and hedgerows, including areas which do not appear to be particularly wet. Adults visit a wide variety of umbellifers, especially Hogweed and Angelica.

See *page 186* for a summary of *Melanogaster*, *Chrysogaster*, *Lejogaster*, *Orthonevra* and *Riponnensia* differentiating features.

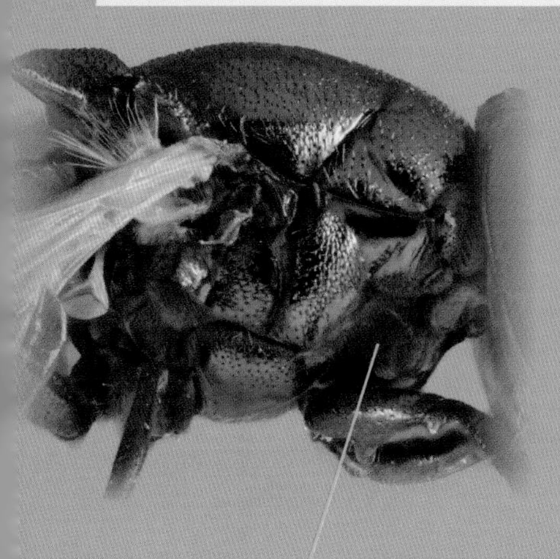

C. cemiteriorum usually has the sides of the thorax grey-dusted just above the front coxae

Chrysogaster cemiteriorum ♀ ×**8**

Chrysogaster species not otherwise covered:

C. virescens – A difficult species to identify. Needs comparison with known specimens of *C. cemiteriorum*. Widespread but very local in the north and west in acidic wetlands, but also occurs in bogs on heathland in southern England.

♂ – *Chrysogaster solstitialis* ×**8** – ♀

Orthonevra

4 British species (1 illustrated)

Small, somewhat metallic hoverflies with a slightly flattened abdomen. Abdomen segment T1 has shiny, metallic edges (like *Riponnensia*), rather than the dull edges of *Melanogaster* (*p. 186*) and *Chrysogaster* (*p. 188*). *Lejogaster* (*p. 192*) has a completely shiny and less flattened abdomen. The presence of a horizontal, dusted band across the face below the antennae is a further useful character. The larvae live among emergent plants at the edges of ponds and ditches.

Orthonevra nobilis 

Frequent

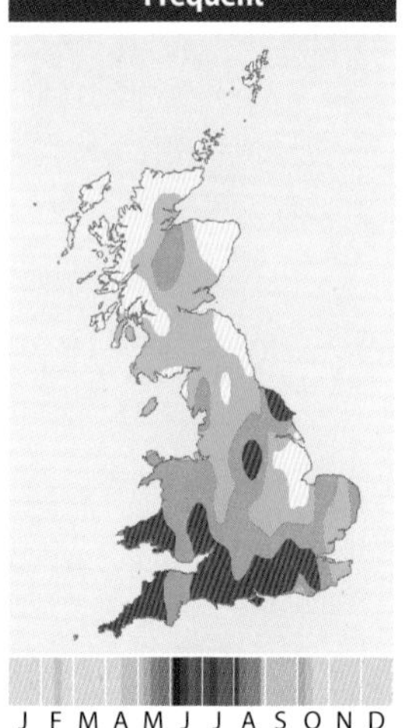

Wing length: 4–5·75 mm

Identification: A small, metallic fly with black legs that is most likely to be confused with *O. geniculata* and *O. intermedia* (which both have partially yellow legs). The third antennal segment is about twice as long as wide and rather pointed (*O. brevicornis* is shorter and blunt). There is often a slight darkening in the centre of the wing.

Similar species: *Orthonevra geniculata* and *O. intermedia* (neither illustrated) and *Lejogaster* species (*p. 192*).

Observation tips: A wetland species which visits a wide range of flowers, especially umbellifers. However, it seems to disperse quite widely and is not infrequently seen at Wild Parsnip and Wild Carrot in drier grasslands.

Riponnensia

1 British species (illustrated)

Formerly placed in *Orthonevra*, with which it shares many characters, such as the abdomen with a flattened matt upper surface and segment T1 with a shiny metallic edge. The larvae live in accumulations of wet, rotting vegetation at the edges of ponds and ditches.

Riponnensia splendens 

Widespread

Wing length: 5·5–7 mm

Identification: The face has a broad, dusted band beneath the antennae (thin in *Orthonevra*). It is also somewhat larger than most *Orthonevra*.

Similar species: *Orthonevra* and *Lejogaster* (*p. 192*),

Observation tips: A wetland species which occurs along pond edges, in marshy places and alongside rivers and canals. It may also be found in wet woodlands. It is a frequent flower visitor, favouring umbellifers but occurring at many others, including Ivy in the Autumn.

See *page 187* for a comparison of *Melanogaster/Chrysogaster*, *Lejogaster* and *Orthonevra/Riponnensia* abdomens and *page 186* for a summary of differentiating features.

Orthonevra: CHRYSOGASTRINI

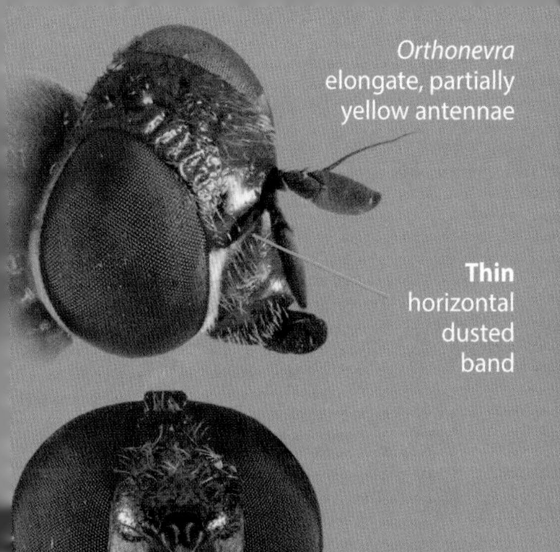

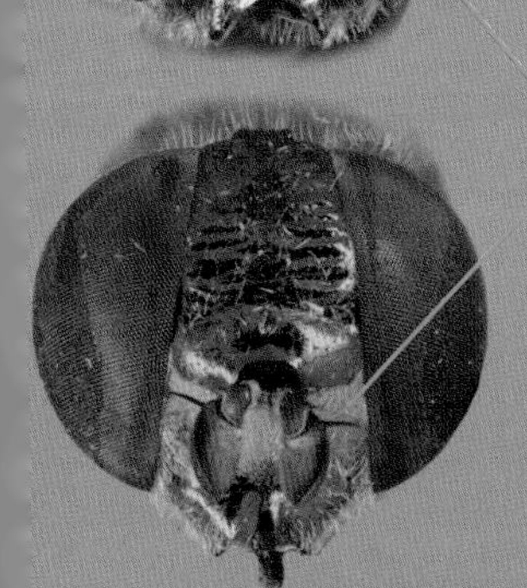

Orthonevra nobilis ♀ ×**8**

Orthonevra species not otherwise covered:

O. brevicornis – A scarce, but widely distributed species in marshes and fens in England and Wales (very few Scottish records).

O. geniculata – A scarce, but widely scattered species in fens and mildly acidic boggy areas.

O. intermedia – Recently discovered at two sites in Cheshire (2006).

Riponnensia: CHRYSOGASTRINI

♀ – *Riponnensia splendens* ×**8** – ♂

Lejogaster

2 British species (both illustrated)

Small, dark hoverflies with a metallic sheen. The eyes are separated in both sexes. There are few other hoverflies with which confusion is likely, the main one being *Riponnensia splendens* (*p. 190*). The larvae are aquatic and live in wet, decaying plant material in the edges of ponds and ditches.

Lejogaster metallina

Widespread

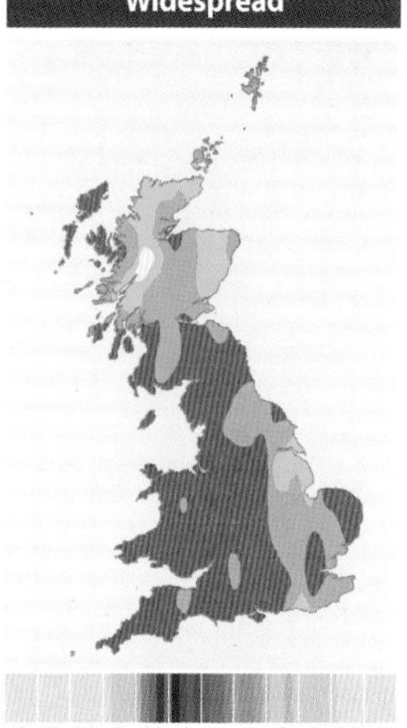

Wing length: 4·75–6·5 mm

Identification: A small, metallic-green hoverfly which has comparatively large antennae. It can only really be confused with *L. tarsata*, which tends to have more bluish reflections and is generally smaller and slimmer. Males have completely black antennae (partially yellow in *L. tarsata*). Females have antennae which are variable in colour, and can be completely black or show some orange/yellow beneath the 3rd segment. They are shorter and rounder in female *L. metallina*; more elongate in *L. tarsata* and always yellowish beneath – usually extensively so.

Observation tips: Found in damp grassland and around wetlands where it visits yellow flowers such as buttercups. It is easily overlooked amongst other hoverflies, particularly species within the tribe Chrysogastrini and *Cheilosia*, and is best found by sweeping. Its near-relative *L. tarsata* tends to be found near the coast or tidal rivers.

Similar species: Both *Lejogaster* are similar. They may be confused with *Orthonevra nobilis*, *O. intermedia* and *Riponnensia splendens* (all *p. 190*), but the abdomen of these three species has a matt upper surface with shiny edges, rather than being wholly metallic. Beware - the very common soldier fly *Chloromyia formosa* (inset) may be mistaken for *Lejogaster* species. Check the wing venation to make sure it is a hoverfly.

Lejogaster tarsata 

Local

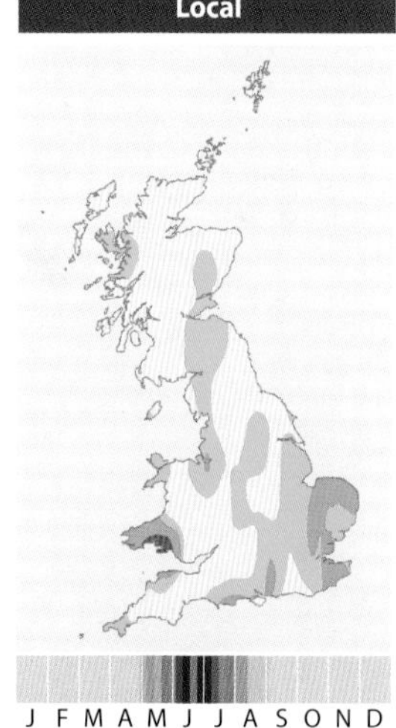

Wing length: 4·5–6 mm

Identification: Slightly smaller and more slender than *L. metallina*. A rather attractive species which tends to have bluish reflections, especially on the thorax of the male and the abdomen of both sexes. The antennae of the male are rather roundish and partially yellow, and those of the female are more elongate than in *L. metallina*.

Observation tips: A rather scarce species compared to *L. metallina* and most often found near the coast, especially on brackish coastal marshes. Can be abundant in some locations.

See *page 187* for a comparison of *Melanogaster/Chrysogaster*, *Lejogaster* and *Orthonevra/Riponnensia* abdomens and *page 186* for a summary of differentiating features.

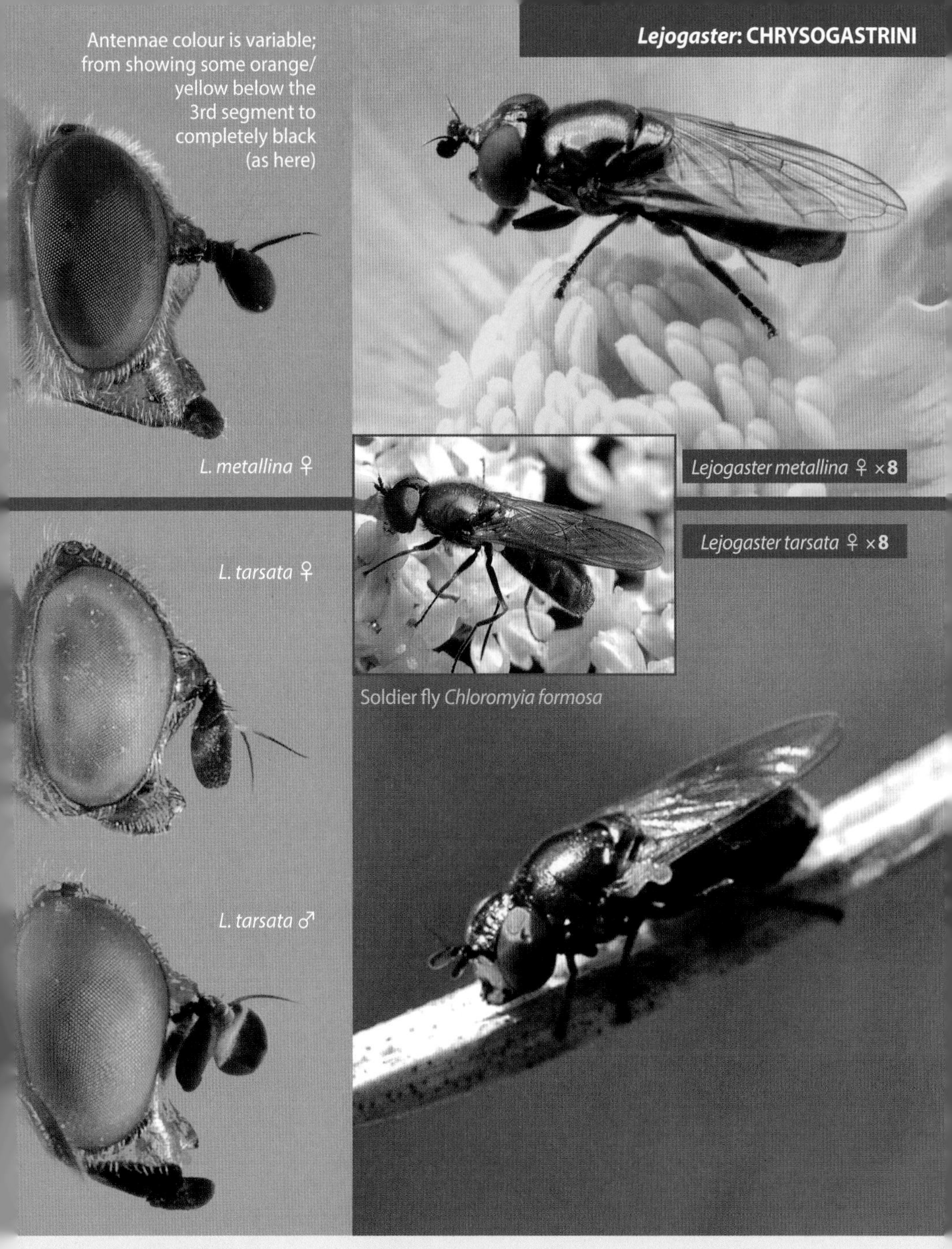

Lejogaster metallina ♀ ×**8**

Lejogaster tarsata ♀ ×**8**

Soldier fly *Chloromyia formosa*

Tips for separating *Lejogaster*:

***L. metallina*:** metallic green; larger and broader than *L. tarsata;*
ANTENNA: ♂ – **black**; ♀ – variable; black or partially yellow; **shorter and rounder** than *L. tarsata*

***L. tarsata*:** metallic bluish-green; smaller and more slender than *L. metallina*;
ANTENNA: ♂ – **partially yellow**; ♀ – partially yellow **more elongate** than *L. metallina*

Brachyopa

4 British species (2 illustrated)

Orange-brown flies which all look very similar and can only be separated by the presence and shape of pits on the antennae. This diagnostic feature requires careful examination under magnification and, consequently, they cannot be identified in the field. There are many additional species in Europe and it is possible that more will be found in Britain. They are often regarded as difficult to find and rare. Adults are usually found hovering near a sap run, resting on the trunk or basking on nearby sunny vegetation. Finding *Brachyopa* is definitely an art, but once the technique is mastered success is more likely. The two species described here are actually quite widespread and are probably more common than records suggest. *Brachyopa* larvae live in sap runs and feed on micro-organisms, such as yeasts, which thrive in the fermenting sap.

Brachyopa insensilis

Wing length: 6·5–7·25 mm

Identification: This is the only British *Brachyopa* that does not have a pit on the inside face of the antennae. The partially grey-dusted scutellum, yellowish humeri (though rather variable and often darker) and almost bare aristae are useful confirmatory features.

Observation tips: This species was thought to be associated with elms and, following Dutch Elm Disease, there was a fear that it would decline. However, *B. insensilis* was subsequently discovered to favour sap runs on Horse-chestnut and has proved to be particularly frequent in urban areas where Horse-chestnuts have been planted as street trees, and in parks and large gardens.

Brachyopa scutellaris

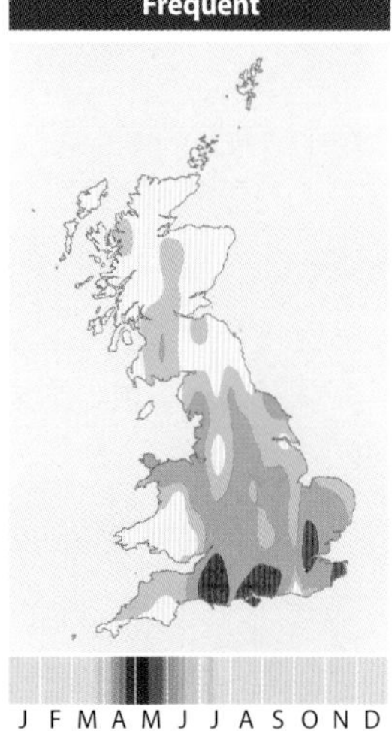

Wing length: 6·5–7·75 mm

Identification: The kidney-shaped pit on the inner face of the antennae is diagnostic, but very careful examination is required as it can vary in shape. The pubescent antennae and yellow humeri are also useful identification features.

Observation tips: Larvae live in sap runs low down on the trunks of oaks and a variety of other deciduous trees (especially Ash, birches and poplars). However, such sap runs are hard to find. Males bask on sunlit leaves, especially the bright green leaves of Sycamore and limes, and will also hover around the sunlit bases of trees. Adults have a short flight period and, on days when they are found, this species may turn out to be surprisingly widespread and common – so it is always worth checking places where you have not found it previously.

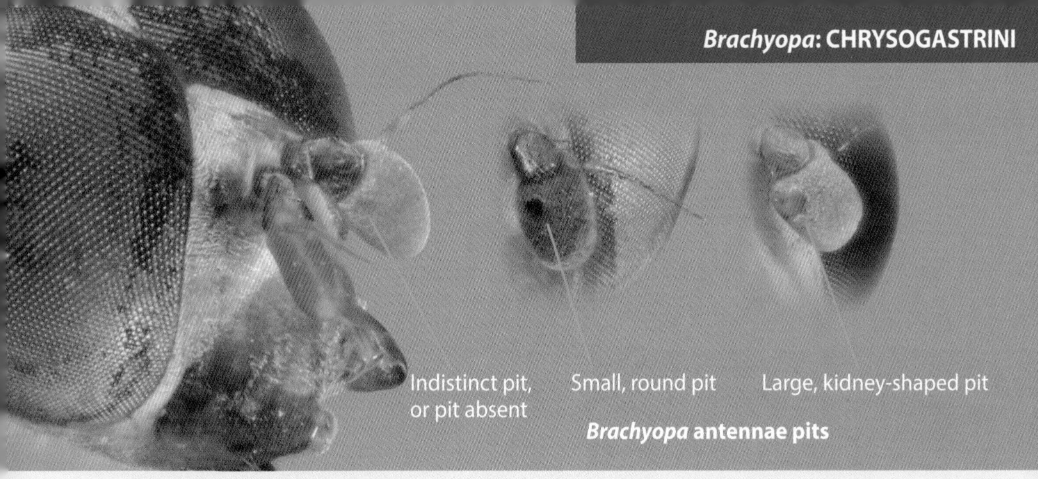

***Brachyopa* antennae pits**

Brachyopa can only be reliably separated by the presence/absence and shape of pits on the antennae. These are indistinct/absent in *B. insensilis*, large and kidney-shaped in *B. scutellaris* and small and rounded in *B. pilosa* and *B. bicolor*. This is a difficult character which needs microscopic examination and known specimens for comparison.

Brachyopa species not otherwise covered:

B. bicolor [NS] – A scarce, forest species of southern England whose range seems to have expanded into the East Midlands and Welsh Borders in recent years.

B. pilosa [NS] – A scarce species with two distinct populations in southern England and northern Scotland. Often, though not exclusively, associated with Aspen and other poplars.

A sap run on Horse-chestnut supporting *B. insensilis* larvae

Brachyopa insensilis ♀ ×**8**

Brachyopa scutellaris ♂ ×**8**

Hammerschmidtia

1 British species (illustrated)

Quite unmistakable – once you realise that it is a hoverfly. In general shape it looks more like a fly from the families Psilidae or Muscidae. Larvae live in rotting sap under the bark of recently fallen, mature Aspen logs. These only provide suitable conditions for a limited period of time, perhaps 2–3 years. Woodlands with sufficient Aspen to ensure a continual supply of suitable fallen timber are now very scarce.

Hammerschmidtia ferruginea

BAP; Endangered

Rare

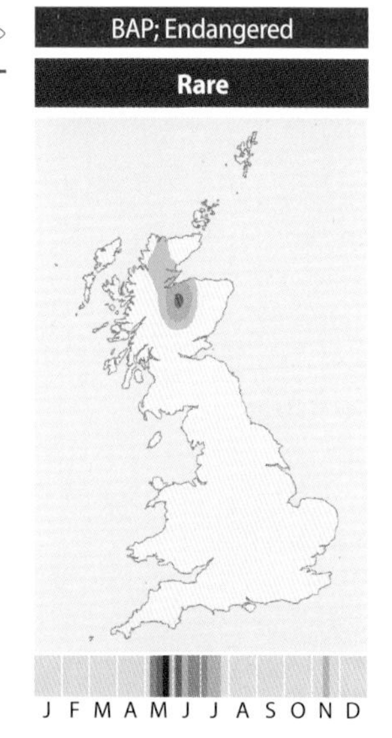

Wing length: 8·25–9·75 mm

Identification: An unmistakable, uniformly red-brown hoverfly; young individuals are pale yellow-brown, becoming considerably darker, even dark brown, with age.

Similar species: *Brachyopa* (*p. 194*) have similar orange-brown abdomens but the thorax is grey. These species do not darken with age to the same degree as *Hammerschmidtia ferruginea*.

Observation tips: An extremely rare hoverfly known from just a handful of Aspen woods in northern Scotland. Recent work has shown that although the population is very restricted, a single log is capable of supporting a surprisingly large number of larvae. These studies also found that adults are capable of dispersing over several kilometres to find new breeding sites.

The abdomen markings of *Myolepta* are distinctive

Myolepta dubia ♂ ×5

Myolepta species not otherwise covered:

Myolepta potens [CR] – Extremely rare. Discovered at two sites in Somerset in the 1940s, but last found there in 1961 and feared extinct. It was rediscovered at Moccas Park NNR, Herefordshire in 2002 and in the Forest of Dean in 2009.

Myolepta

2 British species (1 illustrated)

Easily recognised by the broad yellow margins on abdomen segments T1, T2, and often T3, which create a central black stripe. The larvae live in rot holes in various deciduous trees. *M. potens*, a Critically Endangered and UKBAP priority species, is currently known from only two localities.

Myolepta dubia

Nationally Scarce

Local

Wing length: 6–8·75 mm

Identification: When perched on a flower or sunlit leaf, the yellow markings on the abdomen are covered by the wings and are not obvious, and the impression given is of yellow wing bases. It can easily be overlooked as a somewhat elongate *Cheilosia*, although once the wings are parted the abdominal markings are distinctive. Separation from *M. potens* (not illustrated) can be difficult and needs expert confirmation.

Similar species: The Critically Endangered *Myolepta potens*, *Cheilosia impressa* (*p. 172*) and *Chrysogaster cemiteriorum* (*p. 188*).

Observation tips: A largely southern species in well-wooded situations where there are old trees. Although they are rarely seen, adults visit flowers, especially umbellifers. It is likely that this species would be found more frequently by searching for larvae in rot holes, particularly in Horse-chestnut and Beech.

Guide to Eristalini (and Merodontini: *Merodon*)

The Eristalini are distinctive as their wings have a prominent loop in vein R_{4+5}. Once this feature is learned, the tribe is instantly recognisable. However, this loop also occurs in *Merodon equestris* (Merodontini (*p. 222*)) which is why it is included here. There are 8 genera which form two distinct groups: those with distinct longitudinal stripes on the thorax (*Anasimyia, Helophilus, Lejops* and *Parhelophilus*), and those that are at least vaguely honey-bee mimics (*Eristalis, Eristalinus, Mallota* and *Myathropa*).

1

a) Bumblebee or honey-bee mimic

▶ **2**

b) Not a bumblebee mimic

▶ **3**

2

from **1a**

a) Bumblebee mimic:
Scutellum ground colour black
Hind femur greatly enlarged and with a flange below
Hind legs entirely black

Merodon equestris *p. 222*

Merodon equestris

b) Bumblebee mimic:
Scutellum ground colour yellow
Hind femur normal
Hind tibia partly yellow

Eristalis intricaria *p. 206*

Eristalis intricaria

c) Honey-bee mimic:
Large; with modified hind leg

Eristalis tenax *p. 206*

Mallota cimbiciformis *p. 210*

d) Poor honey-bee mimics:

Most *Eristalis* *pp. 200–207*

Eristalis tenax

3

from **1b**

a) Eyes spotted in life
Scutellum black

Eristalinus p. 208

Beware - the eye spots may disappear in preserved specimens.

b) Thorax with pale bar running across creating a 'Batman' marking towards the back

Myathropa florea p. 210

Beware - these markings are very variable and can be obscure in some Spring specimens which tend to be darker and less colourful.

c) Thorax with pale stripes running lengthways

▶ 4

4

from **3c**

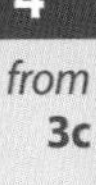

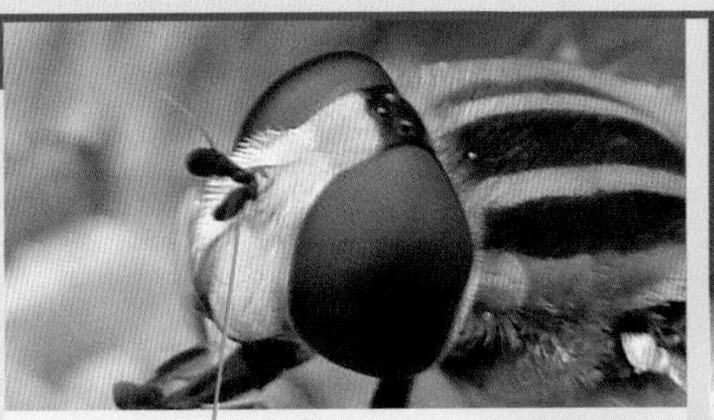

a) Antennae black

b) Antennae orange

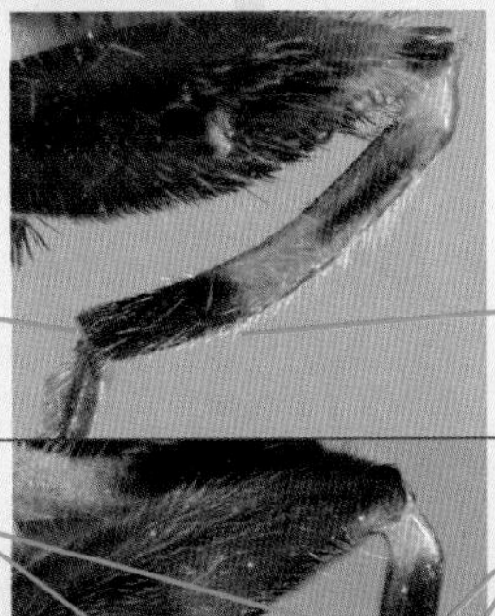

Hind tibia: one dark ring
Helophilus pp. 214–216

Hind tibia: one dark ring
Parhelophilus p. 218

Hind tibia: two dark rings
Lejops p. 216

Hind tibia: two dark rings
Anasimyia p. 212

Eristalis

10 British species (7 illustrated)

This is one of the group of genera that have a loop in vein R_{4+5}. They have a 'petiole' beyond the junction of veins R_1 and R_{2+3} and no eye spots. Often described as bee mimics, but apart from *E. tenax* which is a good honey-bee mimic and the bumblebee mimic *E. intricaria*, they are not very convincing. The larvae are aquatic 'rat-tailed' maggots that live in wet, decaying vegetation, usually in ponds or ditches, but sometimes in farmyard manure pits and silage.

Eristalis pertinax

Wing length: 8·25–12·75 mm

Identification: This is one of the easiest *Eristalis* to identify as the tarsi on the front and middle legs are yellow. A further useful character is the distinctly triangular shape of the abdomen which, especially in males, gives it a particular 'jizz'.

Similar species: Occasionally confused with *Eristalis nemorum*, especially in photographs which do not show the tarsi.

Observation tips: This species occurs almost everywhere, including upland moorland, and throughout the warmer months (March to November). It is one of the first species to appear in the Spring and males characteristically defend territories in woodland rides and around flowering bushes. Remarkably, very few females are seen at this time. It becomes even more abundant in late Summer and can be seen in considerable numbers on the flowers of late-flowering Angelica and thistles until the first frosts.

Tips for field identification of the common *Eristalis* species.

SPECIES	FORM	FACE (*see opposite*)	LEGS	OTHER FEATURES
E. intricaria	**bumblebee mimic; furry**		hind tibia pale on basal half	SCUTELLUM: yellow
E. tenax	**large species**	broad black, shining stripe	hind legs dark and curved with long hairs at the middle of tibia resembling a pollen basket	EYES: each with a vertical band of longer, dark hairs – makes them look subtly striped.
E. pertinax		narrow shining black stripe (set in a broad, dull, darkened area).	front and mid-tarsi yellow-orange	ABDOMEN: rather triangular, especially in male.
E. arbustorum	**small species** (see also *page 204*)	completely dusted	hind metatarsus thickened	WING: stigma diffuse
E. nemorum		narrow black stripe	hind metatarsus not thickened	WING: stigma very discrete, square

If it doesn't match one of these - retain the specimen for more detailed examination!

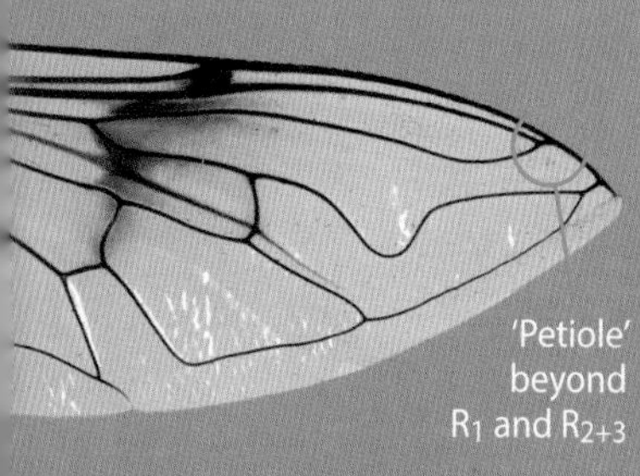

***Eristalis* faces**

E. tenax
broad black shining stripe

E. nemorum
narrow black stripe

E. pertinax
narrow black shining stripe
set in broad, dull darkened area

E. arbustorum
completely dusted

Eristalis species not otherwise covered:

Eristalis abusiva	– See 'Similar species' under *E. arbustorum* (*p. 202*). Widely distributed but uncommon. Most frequent near the coast, especially in the west.
Eristalis cryptarum [CR]	– Illustrated on *page 296*. Extremely rare and restricted to a few localities on Dartmoor. It is not obviously an *Eristalis* at first glance. Once it is recognised as an *Eristalis*, its identity should be obvious because it has completely yellow/orange legs.
Eristalis similis	– A vagrant with 5 British records. A common southern European species which extends northwards in some years and can be frequent on the near continent (*e.g.* the Netherlands, Denmark). A few probably reach our shores at these times.

Eristalis arbustorum

Wing length: 7–10 mm

Identification: A small *Eristalis* with a completely yellow-dusted face with no central shining black stripe. However, as an individual ages, the face can get rubbed allowing black patches to show through – so care is needed. The middle tibiae are pale and strongly darkened near the metatarsal joint. It has a diffuse and extended stigma. It can be confused with *E. nemorum*, which has a sharp-edged stigma, and with the scarcer *E. abusiva* (not illustrated), which has no long hairs on the aristae and a yellow middle tibia (occasionally with diffuse darker smudges near the metatarsal joint).

Similar species: Apart from other small *Eristalis* (see above and table on *page 204*), confusion is most likely with *Epistrophe eligans* (*p. 144*), though the presence of the wing loop should rule this species out.

Observation tips: Widespread and common. Adults occur a long way from breeding sites and visit a wide range of flowers such as knapweeds, ragworts, thistles and umbellifers. Numbers increase through the Spring and it is most abundant in mid-Summer.

Eristalis nemorum

Wing length: 8·25–10·5 mm

Identification: The most straightforward identification feature is the behaviour of the male, which will hover above a female feeding on a flower in a pose which is much-photographed. Both sexes are variable in both size and colouration and identification requires care. The wing stigma is usually discrete and sharp-edged.

Similar species: Most likely to be confused with *Eristalis arbustorum* or *E. abusiva* (not illustrated) with a rubbed face. The wing stigma of other small *Eristalis* is more diffuse. Occasional confusion with *Eristalis pertinax* (*p. 200*) has also been noted – see tables on *pages 200* and *204*.

Observation tips: Widespread, occurring in woodland rides, hedgerows and flowery meadows, often some distance from obvious breeding sites (shallow, nutrient-enriched water). Adults visit a wide variety of flowers. It has a long flight period and can be abundant in mid-Summer.

Eristalis arbustorum ♂ ×**6**
Diffuse wing stigma
Dark hind metatarsus as thick as tibia
Eristalis nemorum ♂ ×**6**
Characteristic behaviour in which the male hovers above the female
Sharp-edged wing stigma
Dark hind metatarsus thinner than tibia

Eristalis horticola

Wing length: 8·25–11·5 mm

Identification: Usually relatively large and more brightly marked than other *Eristalis*, with a prominent dark mark across the centre of each wing (although this is very variable in density and extent). It is most likely to be confused with *E. rupium*, but in *E. horticola* the hind metatarsus is dark.

Similar species: Other members of the genus *Eristalis*, especially in Europe where this genus becomes difficult to separate.

Observation tips: Mainly a woodland and hedgerow species which is widespread, but tends to be more abundant in the north. It is frequently found in association with Bramble and Hawthorn flowers. Males often hover and defend a territory in a similar manner to *E. pertinax*.

Eristalis rupium

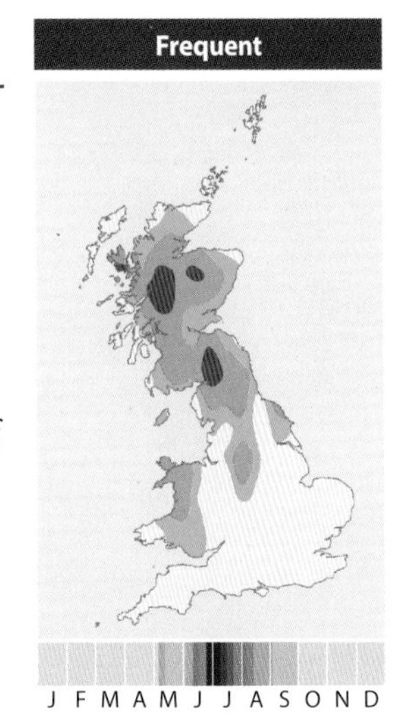

Wing length: 7·75–11·25 mm

Identification: The wings have an almost milky appearance with a dark chocolate marking that is more pronounced than in any other *Eristalis* (beware, however, that newly-emerged individuals lack this marking). The hind metatarsus and the joint with tarsal segment 2 are pale, a character not exhibited by any other *Eristalis*. There are several similar European species that could conceivably be found in Britain (separation of these is based on features of the male genitalia).

Similar species: *Eristalis horticola* in Britain, but there are several similar species in Europe that could potenially turn up.

Observation tips: Mainly an upland species of northern and western Britain. However, in northern Scotland it occurs at low altitudes and has been noted in numbers on Hemlock Water-dropwort flowers along the shore of a sea-loch. Uncommon, but can be abundant where it occurs.

The smaller *Eristalis* species can be extremely tricky. A feature that can be helpful is the colouration and shape of the hind metatarsus (*i.e.* the first segment of the hind tarsus):

<table>
<tr><th>SPECIES</th><th>COLOUR</th><th>HIND METATARSUS (see opposite)</th></tr>
<tr><td>E. arbustorum</td><td>dark</td><td>thickened: at least as thick, or slightly thicker than the tibia</td></tr>
<tr><td>E. nemorum</td><td rowspan="2">dark</td><td rowspan="3">not thickened; thinner than the tibia</td></tr>
<tr><td>E. horticola</td></tr>
<tr><td>E. rupium</td><td>pale</td></tr>
</table>

Eristalis horticola ♂ ×**6**

Eristalis rupium ♀ ×**6**

Eristalis intricaria

Wing length: 8·25–12 mm

Identification: A sexually dimorphic bumblebee mimic with a yellow scutellum. The males are dark with a reddish-brown 'tail', whilst females are somewhat larger and have a white 'tail'.

Similar species: Most likely to be confused with *Cheilosia illustrata* (*p. 162*), but that species lacks a loop in wing vein R_{4+5} and has a black scutellum. Unlike other bumblebee mimics, the hind tibia is half black and half yellow but confusion with *Eriozona syrphoides* (*p. 112*), *Leucozona lucorum* (*p. 112*) and *Volucella pellucens* (*p. 246*) is possible, though these species also lack the wing loop characteristic of *Eristalis*.

Observation tips: Usually found in or near damp places. Males characteristically hover close to nectar sources, such as willows, in the Spring. Both sexes visit a range of low-growing flowers, with a liking for blue and purple ones such as thistles. There are two main peaks in abundance: a small one in the Spring and a much bigger one in mid-Summer.

Eristalis tenax

Wing length: 9·75–13 mm

Identification: There are three straightforward characters that distinguish this large, honey-bee mimic from other *Eristalis*: the eyes have a vertical stripe of longer, dark hairs; the black facial stripe is very wide; and the hind tibia is distinctly enlarged and curved.

Similar species: Could possibly be confused with the rare vagrant *Eristalis similis* (not illustrated), which is strongly dusted on the sides of the thorax.

Observation tips: Although it occurs in the Spring, it is most abundant in late Summer and Autumn and often numerous on Ivy flowers. Females hibernate in sheltered cavities in caves and buildings. A synanthropic species that is now found worldwide. The larvae live in highly enriched aquatic environments including slurry tanks and silage clamps, where they can sometimes occur in vast numbers.

See *Identifying wasp and bee mimics* on *pages 64–66*.

♀ – *Eristalis intricaria* ×**6** – ♂

Eristalis tenax ♀ ×**6**

Eristalinus

2 British species (both illustrated)

The loop in wing vein R_{4+5} and spotted eyes is a combination of characters unique among British hoverflies. The eye spots result from the physical structure of the surface of the eye and are lost in preserved specimens as the eyes dry out. The larvae are of the 'rat-tailed' aquatic type and live among decaying vegetable matter in water.

Eristalinus aeneus

Wing length: 6·25–9·5 mm

Identification: This is a somewhat larger species than *E. sepulchralis* and the absence of hairs on the lower part of the eye is diagnostic. The abdomen lacks any dull patches and is often a slightly bronzy-black in appearance.

Similar species: Apart from *Eristalinus sepulchralis*, there should be no confusion.

Observation tips: An almost entirely coastal species whose larvae live in pools with accumulations of rotting seaweed. Although the main flight period is in mid-Summer, adults occur for much of the year and also hibernate. Widely distributed around the British coast, but most abundant in southern and western locations.

Eristalinus sepulchralis

Wing length: 6·5–8 mm

Identification: Dark, shiny, slightly metallic and rather dumpy hoverflies, with distinct dull patches running along the central axis of the abdomen. The eyes are distinctly spotted and have a full covering of hairs. This species tends to be small compared to the *Eristalis* species with which it is often seen (but the size is quite variable).

Similar species: Apart from *Eristalinus aeneus*, there should be no confusion.

Observation tips: Generally found visiting flowers in the vicinity of nutrient-enriched pools and ditches, for example those affected by agricultural run-off. The larvae live in piles of rotting vegetation in water and also in wet manure. Although widespread across mainland Britain and the Western Isles, this is essentially a southern lowland species. It is much less common than it once was and seems to have undergone a serious decline over the last decade or so.

Eristalinus aeneus ♀ ×**8**

Eristalinus sepulchralis ♀ ×**8**

Mallota

1 British species (illustrated)

One of the genera with a loop in vein R_{4+5} and a very convincing honey-bee mimic. The 'rat-tailed maggot' larvae live in deep, water-filled rot holes in a variety of trees – large, mature or old trees often providing the most suitable breeding sites.

Mallota cimbiciformis

Nationally Scarce

Local

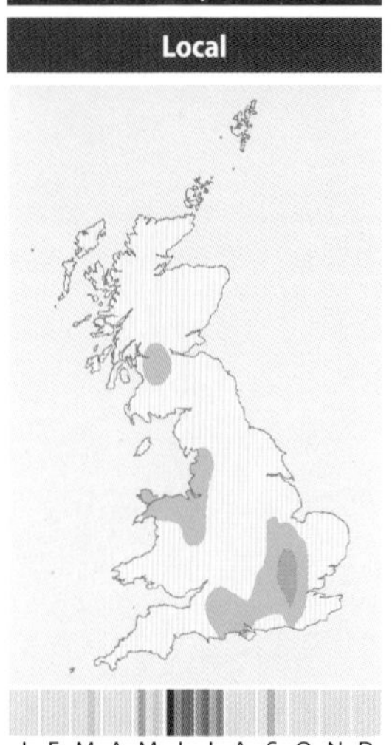

Wing length: 11·25–12·5 mm

Identification: Most likely to be confused with a large *Eristalis* at first sight, but it is actually quite distinct. The most obvious feature is the enlarged hind femur, but the overall furriness and greenish hue are other useful identification features.

Similar species: Other honey-bee mimics such as *Eristalis tenax* (*p. 206*).

Observation tips: Adults are flower visitors which generally stay fairly close to breeding sites. This is a scarce species, with scattered records north to Glasgow. Adults are probably overlooked and it may be easier to find by searching for larvae in water-filled rot holes.

Myathropa

1 British species (illustrated)

A distinctive yellow-and-black species, often described as a wasp mimic, but not very convincing. One of the genera with a loop in vein R_{4+5}; veins R_1 and R_{2+3} reach the wing margin separately. It has a distinctively patterned thorax, with the dark area towards the back resembling the 'Batman' symbol. The 'rat-tailed maggot' larvae live in wet hollows containing decaying leaves and twigs.

Myathropa florea

Widespread

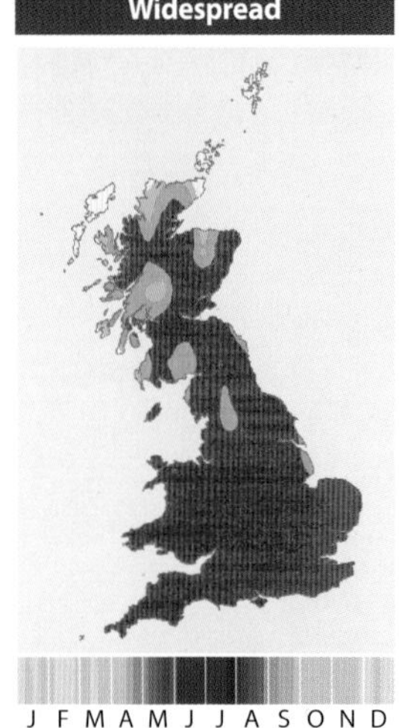

Wing length: 7–12 mm

Identification: As described above, the pattern on the top of the thorax is distinctive. However, although well-marked adults are unmistakable, the markings are very variable and poorly-marked individuals can be difficult to recognise.

Similar species: Poorly-marked *Myathropa* might be initially confused with other Eristalines but should be identifiable.

Observation tips: Widespread and abundant. Although their breeding sites are most often in woodland, they are great opportunists and will breed in anything that holds water – such as a bucket or stray plastic container. Adults occur from April to November and visit a wide range of flowers. They are also frequently seen basking on sunny leaves.

Mallota: ERISTALINI

Mallota cimbiciformis ♀ ×**6**

Myathropa: ERISTALINI

Veins R_1 and R_{2+3} reach the wing margin separately

Distinctive 'Batman' marking on the thorax

♂ – *Myathropa florea* ×**6** – ♀

Anasimyia 5 British species (2 illustrated)

Anasimyia is one of a small group of genera in the tribe Eristalini (including *Helophilus*, *Lejops* and *Parhelophilus* – see table on *page 218*) that have **wings with a loop in vein R_{4+5}** and a **thorax that has pale, longitudinal stripes on the top**. In both the males and females the eyes are separated. They are most similar to *Parhelophilus* (*p. 218*), both having orange antennae, but differ in having two dark rings on the hind tibia, rather than one. *Lejops* (*p. 216*) also has two dark rings on the hind tibia but, as *Helophilus*, differ in their dark antennae. *Anasimyia* species are found in wetlands where the aquatic 'rat-tailed maggot' larvae live among rotting material around the base of emergent plants. Adults fly and rest among water plants, visiting nearby flowers and seldom straying far from breeding sites.

Anasimyia lineata

Frequent

Wing length: 6·25–8·25 mm

Identification: Relatively straightforward to recognise as it is the only British species with the combination of longitudinal stripes on the thorax and a strongly extended face (only *Rhingia* (*p. 178*) has a face shaped like this – but in that genus the feature is even more extreme, and the abdomen is wholly orange.)

Similar species: Two Nationally Scarce species *A. lunulata* and *A. interpuncta* (neither illustrated) are similarly marked but do not have an extended face. The 'comma' markings on abdomen segments T3 and T4 of *A. lunulata* are less angled than those of *A. interpuncta*; the markings of both these species do not approach the 'hockey-stick' look of *A. contracta* or *A. transfuga*.

Observation tips: Look for at the margins of still or slow-flowing water with emergent vegetation, especially Bulrush. Adults will visit a range of flowers nearby. The most widespread and frequent *Anasimyia*, which can be found even in farm and roadside ditches.

Anasimyia contracta

Frequent

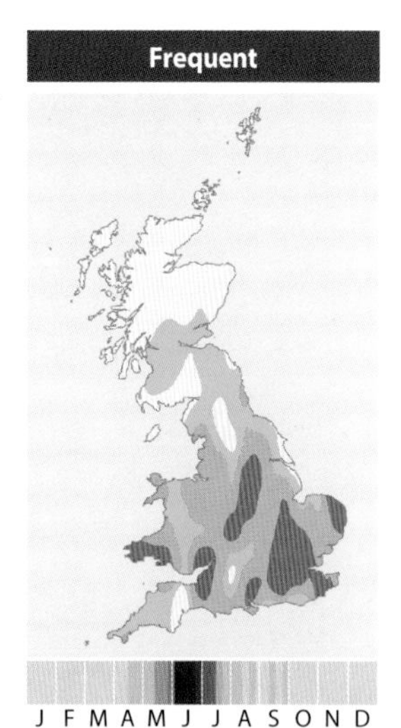

Wing length: 5·0–7·25 mm

Identification: This is perhaps the most likely *Anasimyia* species to be encountered. It has a somewhat 'wasp-waisted' abdomen, with a narrower than long segment T2 and greyish, 'hockey-stick'-shaped markings.

Similar species: The abdomen of *A. transfuga* (not illustrated) has similar markings but with segment T2 parallel-sided and wider than long.

Observation tips: Look for in emergent vegetation at the margins of still water. Adults visit buttercups and other flowers nearby. Larvae have been found living between the leaf sheaths of rotting plants. Widely distributed, but local.

The strongly extended face of *A. lineata* is diagnostic within the Eristalini

Anasimyia lineata ♂ ×**6**

♂ – *Anasimyia contracta* ×**6** – ♀

Anasimyia species not otherwise covered:

A. interpuncta [NS] – A rare species largely found in the larger river valleys of East Anglia, including the Norfolk Broads, and in the Thames Marshes. Recently discovered in the Somerset Levels.

A. lunulata [NS] – A scarce species largely found in the west, especially west Wales, where it occurs in peatlands.

A. transfuga – Like *A. contracta*, this species has hooked abdominal markings, but T2 is not narrowed. It is much less common than *A. contracta*, but widely distributed in Wales and the southern half of England. It tends to occur in more shaded situations than other members of the genus.

Helophilus

5 British species (3 illustrated)

Helophilus is one of a small group of genera (including *Anasimyia*, (*p.212*) *Lejops* (*p.216*) and *Parhelophilus* (*p.218*) – *see table on page 218*) that have **wings with a loop in vein R_{4+5}** and a **thorax that has pale, longitudinal stripes on the top**. Like *Lejops*, they have dark antennae, but the hind tibia only has one dark ring, whereas *Lejops* has two. The larvae are aquatic 'rat-tailed maggots' found in wet, decaying vegetation in the edges of ponds and ditches.
The sex of the adult is important in identification and, as the eyes are separated on the top of the head in both sexes, it is necessary to check the shape of the end of the abdomen to tell them apart (genital capsule present in the male). There are three common species; the other two, *H. groenlandicus* and *H. affinis* are thought to be very rare vagrants.

Helophilus hybridus

Wing length: 8·5–11·25 mm

Identification: Like *H. pendulus*, the face is yellow with a central black stripe. Males are easily separated from *H. pendulus* because they lack a black band separating the yellow markings on abdomen segments T2 and T3. Females are more difficult: more than half of the hind tibia is black (only about one third in *H. pendulus*) and the yellow dusting on the frons extends right back to the ocellar triangle.

Observation tips: A wetland species which tends to be found near its breeding habitat, where it visits a variety of flowers. It occurs widely across Britain, but is less common in the north.

Similar species: All *Helophilus* are similar. *H. trivittatus* can be distinguished by its yellow face (*p.216*). *H. groenlandicus* and *H. affinis* (neither illustrated) are unlikely to be encountered as they are considered to be rare vagrants.

Helophilus pendulus

Wing length: 8·5–11·25 mm

Identification: Like *H. hybridus* the face is yellow with a central black stripe. Males are easily separated from that species because the yellow markings on abdomen segments T2 and T3 are separated by a black band. Females are more difficult: only about one third of the hind tibia is black (over half in *H. hybridus*) and the yellow dusting on the frons stops abruptly, well in front of the ocellar triangle.

Observation tips: A common and widespread species which visits a wide range of flowers in a variety of habitats, including gardens, which may be far from breeding sites.

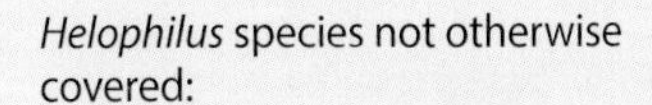

Helophilus species not otherwise covered:

H. affinis – Vagrant. Single record from Fair Isle in 1982.

H. groenlandicus [DD] – Very few records from north west of Scotland, mainly from islands of the inner Hebrides. This is a species of tundra lakes, also found at high altitude in some European mountains. Its status in Scotland has not been established. Rotheray & Gilbert (2011) regard it as a vagrant from the Arctic, but it may be resident.

Helophilus hybridus ♂ ×**5**

♀ – *Helophilus pendulus* ×**5** – ♂

Helophilus trivittatus

Widespread

Wing length: 10·25–12·25 mm

Identification: Unlike the other two common *Helophilus* species, the face does not have a central black stripe (although there is sometimes a reddish-brown patch around the 'nose' in the middle of the face). It is also noticeably larger than the others and tends to be a brighter, more lemon-yellow colour with more extensive yellow markings on the abdomen.

Similar species: Large *Helophilus pendulus* or female *H. hybridus* (both *p. 214*) if viewed from above; however, once the face is seen, any possible confusion should be eliminated.

Observation tips: A migratory species which is most frequently encountered near the coast or along major river courses. It visits flowers in a wide variety of habitats, including in dry places that are remote from any obvious breeding sites.

Lejops

1 British species (illustrated)

Lejops is one of a small group of genera (including *Anasimyia* (*p. 212*), *Helophilus* (*pp. 214–216*) and *Parhelophilus* (*p. 218*) – see table on *page 218*) that have **wings with a loop in vein R_{4+5}** and a **thorax that has pale, longitudinal stripes on the top**. Like *Helophilus*, they have dark antennae, but the hind tibia has two dark rings, whereas *Helophilus* has only one. The larvae are unknown, but are assumed to live in the wet, decaying vegetation at the base of Sea Club-rush beds where the adults are found.

Lejops vittatus

Wing length: 8–10 mm

Identification: Reminiscent of an *Anasimyia* but has black antennae, rather than orange. *Lejops* appears narrower than members of the other genera that have longitudinal stripes on the thorax. The two dark rings on the legs should readily separate it from *Helophilus*.

Similar species: *Anasimyia*, *Helophilus* and *Parhelophilus* (see table on *page 218*).

Observation tips: A rare species of coastal grazing marshes where it is found in and around beds of Sea Club-rush. Confined to a few coastal marshes including the Broads, the Thames marshes, the Somerset Levels and the Gwent Levels.

Helophilus: ERISTALINI

Helophilus trivittatus ♀ ×**6**

Lejops: ERISTALINI

Lejops vittatus ♀ ×**6**

Parhelophilus

3 British species (1 illustrated)

One of a small group of genera (including *Anasimyia* (*p.212*), *Helophilus* (*pp.214–216*) and *Lejops* (*p.216*) – *see table below*) that have **wings with a loop in vein R_{4+5}** and a **thorax that has pale, longitudinal stripes on the top**. Like *Anasimyia*, they have orange antennae, but their hind tibia has only one dark ring whereas *Anasimyia* has two. The two commoner species (*P. frutetorum* and *P. versicolor*) have a general orange appearance that is quite distinctive. The 'rat-tailed maggot' larvae live in decaying plant material at the base of emergent vegetation, especially Bulrush, in ponds and ditches.

Parhelophilus frutetorum / versicolor 

Frequent

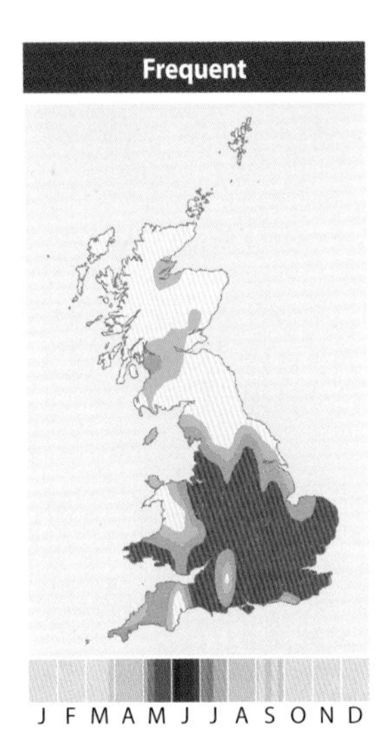

Wing length: 7–9 mm

Identification: These two species are very difficult to separate, especially the females. Determining the sex is also difficult because the eyes of the male do not meet on top of the head – the best way to check is to look for a genital capsule; the abdomen of females is tapered. Male *P. frutetorum* have a small tubercle on the underside of the hind femur, a feature which is lacking in *P. versicolor*.

Similar species: The Nationally Scarce *Parhelophilus consimilis* (not illustrated) bears some resemblance to these species but could be confused in the field with some *Anasimyia*. *Anasimyia* (*p.212*) and *Helophilus* (*pp.214–216*) – see table *below*.

Observation tips: Around emergent vegetation in ponds and ditches, especially Bulrush; they are regular flower visitors and can be seen at Yellow Iris. *P. frutetorum* seems to be less faithful to breeding sites and can be found some distance away. The map shows the combined distribution of the two species.

Differentiating features of the Eristalini that have a striped thorax.

GENUS	ANTENNAE	HIND TIBIA
Anasimyia *p.212*	Orange	**two** dark rings
Parhelophilus *above*		**one** dark ring
Helophilus *pp.214–216*	Black	**one** dark ring
Lejops *p.216*		**two** dark rings

Parhelophilus frutetorum ♂ ×**6**

<u>*Parhelophilus* species not otherwise covered:</u>

Apart from male *P. frutetorum* (where the tubercle underneath the hind femur confirms its identity upon close examination), *Parhelophilus* are difficult to separate, especially in the females. Comparison with known specimens is necessary for accurate identification.

P. consimilis [NS] – This species tends to look rather darker and yellower than the other two and is perhaps more like an *Anasimyia* in the field. It is rare but widely distributed, in poor-fen habitats primarily in Wales and as far north as southern Scotland.

♂ *P. frutetorum* has a hind femur with a curious, long-haired tubercle

Guide to Merodontini

This tribe comprises three genera, two of which contain a single species that are very different from each other in appearance. **See the respective species accounts for more information.**

Eumerus

5 British species (4 illustrated)

Small, dark hoverflies with distinct hair bands on the abdomen and enlarged hind femora. The shape of the re-entrant outer cross-vein is distinctive. The larvae are bulb and rhizome dwellers and at least one species has been introduced to Britain in imported bulbs. Many more species occur in Europe and accidental introductions are always possible.

Eumerus ornatus

Wing length: 4·5–6·25 mm

Identification: Very similar to other *Eumerus* but generally a little larger than most *E. funeralis/strigatus.* The ocelli lie considerably farther forward on the frons than other similar species but this is a feature that causes considerable confusion. Careful examination is needed to be certain and records are generally not accepted without voucher specimens.

Similar species: Confusion with *E. funeralis/sogdianus/strigatus* is most likely and careful examination of the position of the ocelli on the frons is necessary.

Observation tips: This species usually flies close to the ground in dappled light in woodlands. On several occasions it has been found along wet rides with stands of Meadowsweet.

Eumerus funeralis / strigatus

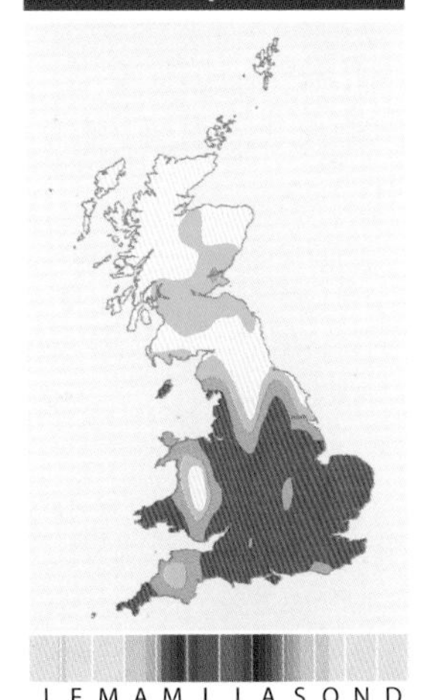

Wing length: 3·5–6·25 mm

Identification: Two very similar species that can only be separated by very close examination of the distribution of hair on the underside of the hind femora; *E. funeralis* has a distinct bare area. Females of *E. strigatus* also have dusting on the top of the head which is much more extensive than in *E. funeralis.*

Similar species: *E. sogdianus* (not illustrated), which can only be separated from *E. strigatus* on the basis of the male genitalia. Females cannot be separated from those of *E. strigatus.* *E. ornatus* is very similar but the ocellar triangle is located farther forward on the frons – a character best appreciated once a specimen has been studied (see *opposite*).

Observation tips: *E. funeralis* is found in gardens, including those in urban areas, and open countryside, visiting a wide range of low-growing flowers. The larvae feed in daffodil and other bulbs where rot is present. *E. strigatus* mainly occurs in open countryside and is a regular visitor to umbellifers. Its larvae have been found in a variety of bulbs and also Iris rhizomes.

E. ornatus

Eumerus ornatus ♂ ×**10**

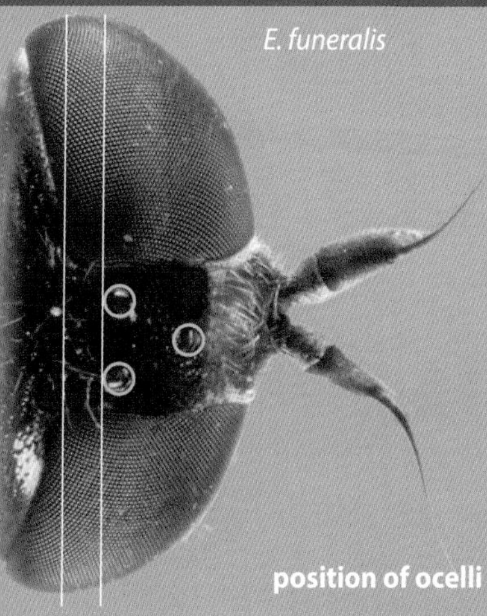
E. funeralis

position of ocelli

The ocelli of *E. ornatus* are positioned farther forward than those of similar *Eumerus* – although this feature requires careful study

Eumerus funeralis ♀ ▲ ♂ ▼ ×**10**

Eumerus species not otherwise covered:

E. sogdianus – Recently added to the British list from two males found on the Isle of Wight.

hind femora

E. funeralis
bare, polished patch

E. strigatus
all dull

Eumerus sabulonum

Wing length: 3·25–5·25 mm

Identification: This should be the most straightforward *Eumerus* to identify because the abdomen has reddish markings. A useful pointer, although not conclusive, is that it is predominantly a coastal species.

Similar species: No other *Eumerus* in Britain has red markings on the abdomen.

Observation tips: A scarce species which can be found on cliffs and sand dunes along the west coast of England, Wales and into southern Scotland. The larvae have been found to feed in the roots of Sheep's-bit in Denmark and adults will also visit this plant.

Merodon

1 British species (illustrated)

A moderately-sized polymorphic bumblebee mimic with a loop in vein R_{4+5}. The only other bumblebee mimic with a wing like this is *Eristalis intricaria* (*p. 206*), but that species has partly yellow hind legs (wholly black in *Merodon*). Larvae develop in bulbs, especially daffodils, but it has been recorded from a wide variety of other bulb-forming plants. Many more *Merodon* species occur in Europe, especially around the eastern Mediterranean, and it is quite possible that others could turn up in Britain as accidental imports.

Merodon equestris

Wing length: 8·5–10·25 mm

Identification: In addition to the features mentioned above, the hind femora are swollen and have a distinct triangular projection towards the apex. A variety of colour forms have been recognised which mimic different species of bumblebee (see *page 31*).

Similar species: Other bumblebee mimics such as *Eriozona syrphoides* (*p. 112*) and *Eristalis intricaria* (*p. 206*). Photographs of *Eristalis tenax* (*p. 206*) have been known to be labelled erroneously as a form of this species, but careful checking of leg colour and the presence of a projection on the hind femora should eliminate any possibility of confusion.

Observation tips: Believed to have been introduced into Britain in daffodil bulbs imported from Europe around the end of the 19th Century. Adults are frequent in gardens and urban areas across the country, but also occur in the wider countryside.

See *Identifying wasp and bee mimics* on *pages 64–66*.

Eumerus: MERODONTINI

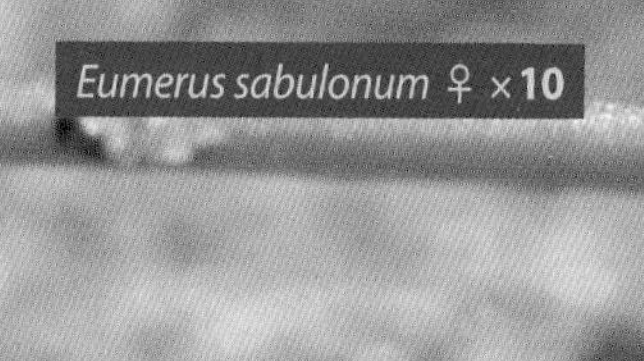

Eumerus sabulonum ♀ ×**10**

The reddish-base to the abdomen separates *E. sabulonum* from other *Eumerus* species

Merodon: MERODONTINI

A variety of colour forms occur (see *page 31*)

The triangular flange on the thickened hind femora is unique amongst the bumblebee mimics

Merodon equestris ×**6**

Psilota

1 British species (illustrated)

Related to *Merodon* and *Eumerus* but looks very different. It is a steely blue-black hoverfly with bright red eyes, which is more like one of the blue-black muscid flies (*e.g. Hydrotaea*) or the hoverfly *Cheilosia pagana*. **It is also the only British hoverfly without any trace of a vena spuria**. Larval biology is unclear, but is probably associated with rotting wood in old trees in some way.

Psilota anthracina

Nationally Scarce

Scarce

Wing length: 6–7·5 mm

Identification: Fairly distinctive – once you realise that it is a hoverfly. The face is flat with a slightly jutting mouth margin. The wing base and stigma are distinctly yellow.

Similar species: *Cheilosia* – especially *C. pagana* (*p. 166*) (which has indications of silvery patches under certain reflected light), *C. impressa* (*p. 172*) and *Platycheirus albimanus* (*p. 80*), though the presence of a vena spuria in these species will rule out *Psilota*.

Observation tips: A rare, southern species most frequently found in woodlands with old trees, particularly in the New Forest and Windsor Great Park, but also in the Midlands. It is a flower visitor, especially to Hawthorn blossom, with a rather brief flight period in May to early June.

A blue-black muscid fly, *Hydrotaea aenescens*, with which *Psilota* could be confused.

Psilota anthracina ♂ × **10**

Guide to Pelecocerini

A single genus, *Pelecocera*, **easily recognised by the very distinctive large, half-moon shaped antennae and thickened arista**.

Pelecocera

3 British species (2 illustrated)

These are small but highly distinctive hoverflies with unusual half-moon shaped antennae that are rather large in proportion to the face – once you look closely. They are small and rather inconspicuous yellow-and-black hoverflies which occur in heathy habitats where they often visit low-growing flowers, and are therefore likely to be overlooked. The larvae are unknown.

Pelecocera scaevoides 

Wing length: 4·0–6·25 mm

Identification: This is a tricky species to separate from the much rarer *P. caledonicus*, in which the yellow on the abdomen is infused with grey dusting and the bristle immediately in front of the wing base is absent (present in *P. scaevoides*).

Similar species: *Pelecocera caledonicus* (see above) but may be overlooked as a *Melanostoma* (*p. 76*) or a *Platycheirus* (*pp. 78–91*) in the field, until the shape of the antennae is seen. *P. tricincta* should not cause confusion as this is restricted to southern England.

Observation tips: Rather scarce and confined to Caledonian pinewoods and mature conifer plantations in central and northern Scotland, where it can be found visiting Tormentil and other yellow flowers along paths and rides. The Vulnerable *P. caledonicus* has a similar distribution.

Pelecocera tricincta 

Wing length: 3·5–5·25 mm

Identification: Readily recognised by the unusual shape of the antennae; there should be no confusion with *P. scaevoides* which are confined to northern Scotland.

Similar species: Can be overlooked as *Melanostoma* (*p. 76*) or *Platycheirus* (*pp. 78–91*).

Observation tips: A southern heathland species which is normally found visiting yellow flowers such as Tormentil and hawkweeds in the grassy margins of tracks and paths. Found on the heaths of Dorset, Hampshire and Surrey, and less commonly in Devon and West Sussex. It can be reasonably common within its limited range.

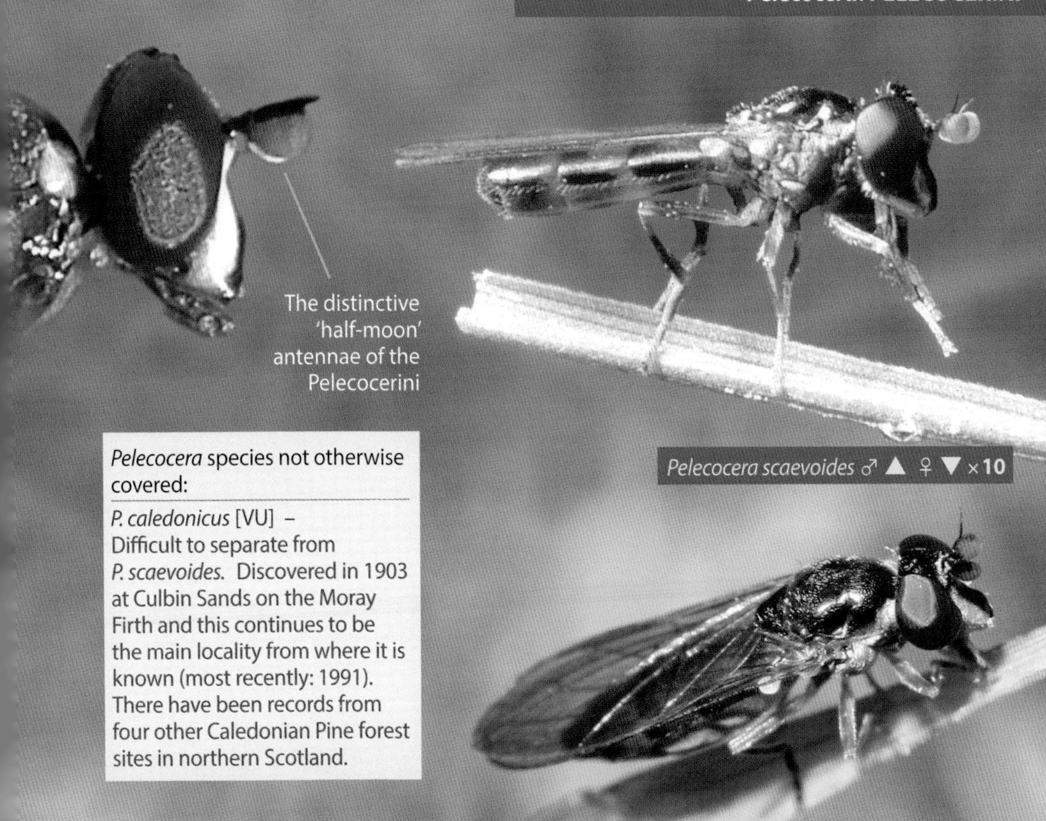

Pelecocera species not otherwise covered:

P. caledonicus [VU] – Difficult to separate from *P. scaevoides*. Discovered in 1903 at Culbin Sands on the Moray Firth and this continues to be the main locality from where it is known (most recently: 1991). There have been records from four other Caledonian Pine forest sites in northern Scotland.

Pelecocera scaevoides ♂ ▲ ♀ ▼ ×**10**

Pelecocera tricincta ♂ ×**10**

Guide to Pipizini

The five genera within this tribe are predominantly black though some bear round, yellow markings on abdomen segment T2. Many species are small to medium-sized and sit in a distinctive manner on leaves and flowers with their wings looking somewhat delta-shaped. The humeri are hairy and in most cases this is readily apparent. **The most useful character for separating this tribe from all others is the flat face, covered in long, drooping hairs, with no evidence of a central prominence and without a distinctly projecting mouth margin**. It is a tribe that is frequently overlooked by recorders; not least because identification of the species can be very challenging. Pipizines are rarely identifiable from photographs with any certainty and all identifications should be treated with caution!

Pipiza austriaca

Pipizini face

The flat face, covered in drooping hairs, with no central prominence and lack of distinctly projecting mouth margin typifies members of the Pipizini.

Psilota anthracina

Beware – *Psilota anthracina* [Merodontini] (*p. 224*) may also key out here. It is a shining bluish-black hoverfly that has a flat face with a strongly pointed mouth-edge. However, *Psilota* lacks a vena spuria (see *page 56*) so differentiation from a pipizine hoverfly should be straightforward. *Psilota* is easily confused with some blue-black muscid flies and so great care should be taken with its identification.

1

a) Abdomen with only two obviously visible segments (T4 is not apparent)

Triglyphus primus *p. 238*

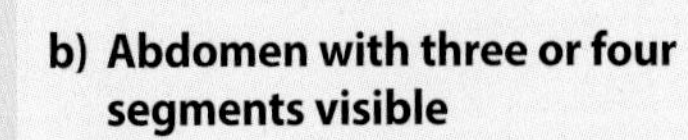

b) Abdomen with three or four segments visible

▶ 2

2

from **1b**

a) Upper-outer cross-vein of wing curved and more upright

Trichopsomyia females have yellow spots on abdomen segment T2; otherwise these are small and entirely black hoverflies.

Legs with pale hairs

Pipizella *p. 236*

Hind tibia with black hairs

Trichopsomyia *p. 238*

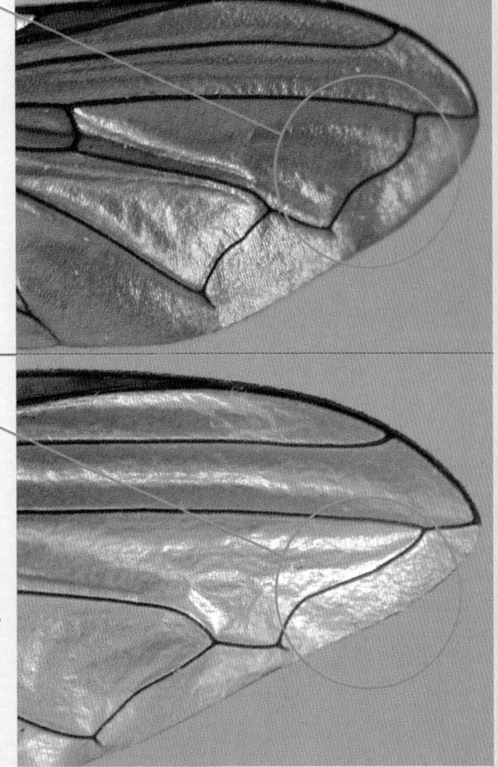

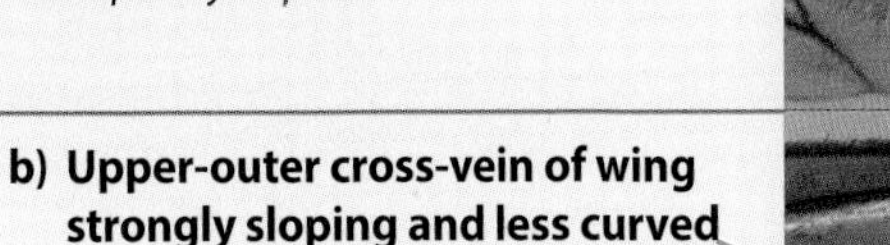

b) Upper-outer cross-vein of wing strongly sloping and less curved

In general, *Pipiza* are larger (*Heringia* tend to have a wing length less than 6 mm), but small *Pipiza* do occur. Males of the sub-genus *Neocnemodon* (within *Heringia*) have a distinctive spur on their hind trochanter (see *page 235*). *Pipiza* often have yellow spots on the abdomen.

Pipiza *pp. 230–233*

Heringia *p. 234*

See accounts for more information on identification of these difficult genera.

Pipiza

7 British species (3 illustrated)

Medium-sized, black hoverflies with a flat, hairy face, and often with a wing cloud. Many of the species have yellow spots on the abdomen segment T2. **However, they are very variable and, in many cases, it is difficult to be absolutely certain of their identity**. Indeed, of all the genera within the family Syrphidae, this is regarded as one of the most intractable. For this reason, only the more readily identifiable species are dealt with here. The larvae feed on aphids, often in enclosed situations such as aphid-induced galls and leaf curls. A major revision of the genus Pipiza in 2013, including DNA analysis, resulted in some changes to species names (*P. bimaculata* is now *P. notata* and *P. fenestrata* is now *P. fasciata*), but the overall composition of the British fauna remains the same.

Pipiza austriaca

Wing length: 6–8 mm

Identification: Amongst the largest species of the genus and fairly easy to recognise because it has ridges on the underside of the hind femora, towards the apex, which gives them a thickened appearance. The abdomen is black in both sexes (any apparent markings are due to bands of hair, not coloured spots), and the wings have strong but diffuse black clouds.

Similar species: Most likely to be confused with *Pipiza noctiluca* (*p.* 232) and *P. fasciata* (not illustrated), but possibly also with the genera *Pipizella* and *Heringia* (see table *below*).

Observation tips: Frequently found visiting flowers such as buttercup and Hogweed in meadows, hedgerows and woodland rides and edges. Widespread in England and Wales, north to the Scottish lowlands, but not usually abundant.

Differentiating features of the Pipizini.

GENUS	MAIN FEATURE	OTHER FEATURES
Pipiza *pp.* 230–233	WING: Upper-outer cross-vein strongly sloping and less curved (see *page* 229)	HEAD: ♀ triangular dust bars; ♂ frons less inflated (see *page* 235)
Heringia *p.* 234		HEAD: ♀ narrow dust bars; absent in sub-genus *Neocnemodon*; ♂ frons more inflated (see *page* 235)
Pipizella *p.* 236	WING: Upper-outer cross-vein curved and more upright (see *page* 229)	LEGS: **pale hairs** ABDOMEN: ♀ wholly black
Trichopsomyia *p.* 238		LEGS: **black hair on hind tibia** ABDOMEN: ♀ yellow spots on T2
Triglyphus *p.* 238	ABDOMEN: **2 segments visible**	

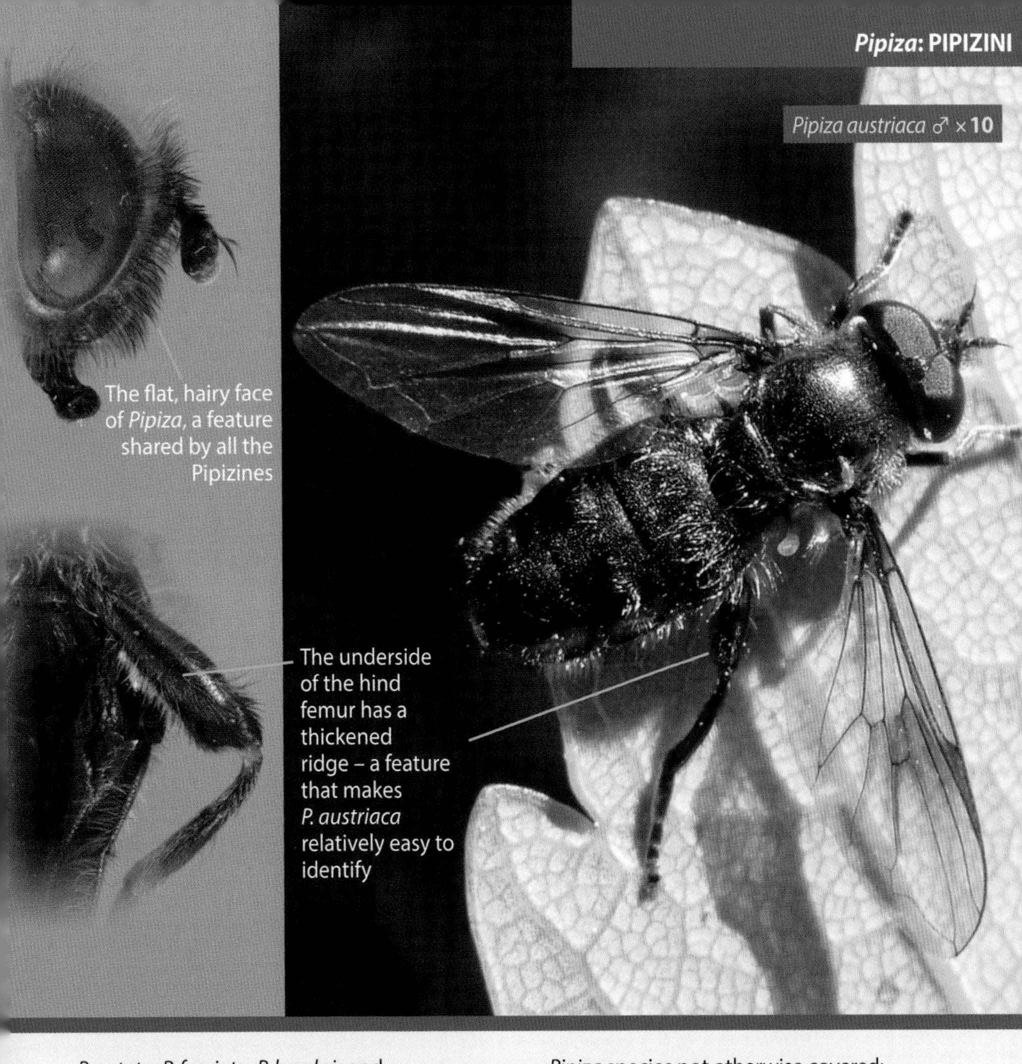

P. notata, *P. fasciata*, *P. lugubris* and ***P. noctiluca*** are regarded as part of an extremely difficult complex of species and are avoided by many recorders. Consequently, their distribution and status are not very well established.

Identification of Pipiza is based upon a range of characters:

Most critically the distribution of different coloured hairs on the face, tergites 4 and 5, and margins of the tergites.

The degree of darkening of the wings is also used, but requires comparison with voucher specimens as it is open to serious misinterpretation. DNA analysis has also been used and although this has resolved a few issues, it has also raised others.

Pipiza species not otherwise covered:

P. fasciata – A few, mostly old, records mostly from south of the Humber, but with a scatter north to Fife. Difficult to distinguish from *P. noctiluca* (*p.232*) but often larger.

P. festiva – Single, Irish record.

P. lugubris [NS] – Widely distributed across England and Wales north to the Humber, but mostly southern central England.

P. notata – Widely scattered, most frequent in southern England.

Pipiza luteitarsis

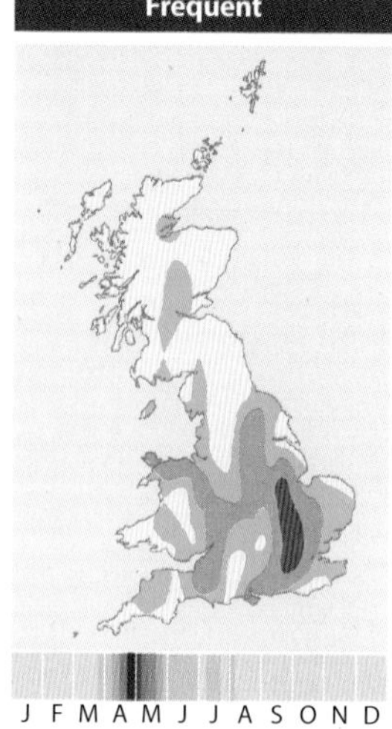

Wing length: 6·5–8 mm

Identification: Most *Pipiza* have some degree of yellow on the tarsi, but *P. luteitarsis* has extensively yellow tarsi on the front and middle legs (though occasionally the outermost tarsal segment is darkened). Both sexes have yellow spots on abdomen segment T2 , but they are smaller in males and sometimes rather vague.

Similar species: Most likely to be confused with *Pipiza noctiluca* and *P. fasciata* (not illustrated) but possibly also with the genera *Pipizella* and *Heringia* (see table on *page 230*).

Observation tips: Most frequently found around elm trees, where the larvae feed upon aphids, which causes the leaves to curl. The adults bask on sunlit leaves and males hover close to the foliage, defending sun sp0ots. An uncommon, predominantly southern species, with few Scottish records.

Pipiza noctiluca

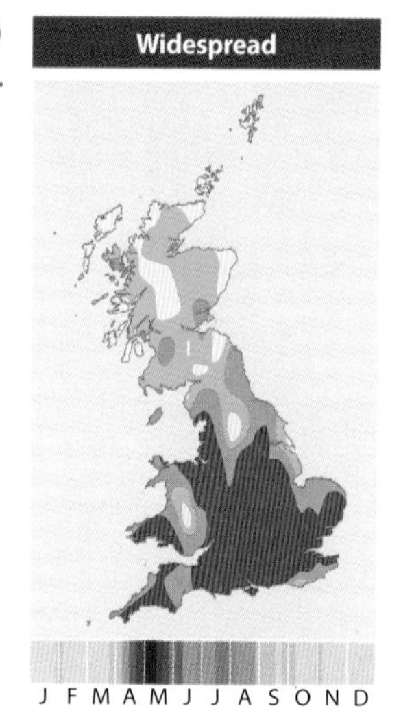

Wing length: 6·5–8 mm

Identification: An enormously variable species which may prove to be a complex of several separate species. The size, intensity of wing shading and tarsal colouration (from largely dark to having yellow tarsi on the middle leg) all vary. The female abdomen typically has a pair of yellow spots, whilst the male is usually unmarked, but the extent of spotting also varies greatly and several different forms have been described (a male with yellow spots on T2 is illustrated).

Similar species: Most likely to be confused with *P. fasciata* (not illustrated) and *Pipiza luteitarsis* but possibly also with the genera *Pipizella* and *Heringia* (see table on *page 230*).

Observation tips: Widespread and fairly common throughout most of southern Britain, as far north as about the Lake District, but scarcer further north. A regular flower visitor, especially to umbellifers, but equally likely to be found basking on sunlit leaves.

Pipiza luteitarsis ♂ ×**10**

Pipiza noctiluca ♂ ×**10**

Heringia

7 British species (3 illustrated)

Small, black hoverflies that have flattened, hairy faces and modifications on the legs of some males. There are two sub-genera (*Heringia* with two species and *Neocnemodon* with five). **All these species, and the majority of the others within the tribe Pipizini, require microscopic examination to confirm identification**. The larvae feed on aphids.

Heringia heringi

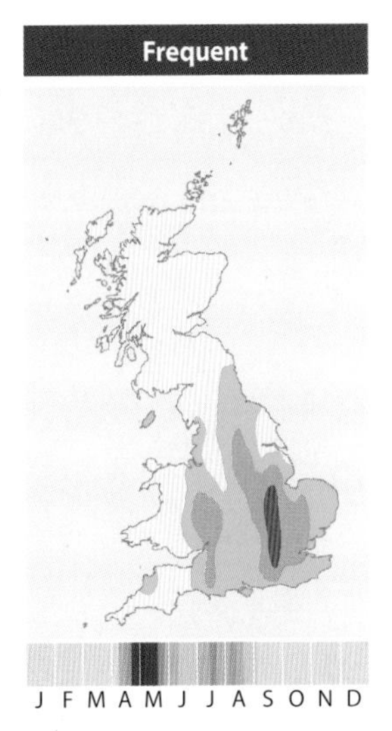

Wing length: 5·5–6·25 mm

Identification: This and *H. senilis* can only be separated by microscopic examination of their genitalia. Males lack the spine that projects from the trochanter in members of the sub-genus *Neocnemodon*. Females have narrow dust bars on the face which are absent in *Neocnemodon*.

Observation tips: The larvae feed on aphids within galls on poplars, elms and willows. Adults are most likely to be found basking on sunlit leaves and do not usually visit flowers.

Similar species: Other members of the genus *Heringia* and some small *Pipiza*, of which the males have a less inflated frons and no spurs or leg modifications and females have triangular dust bars (see table on *page 230*).

Heringia pubescens[1] / *vitripennis*

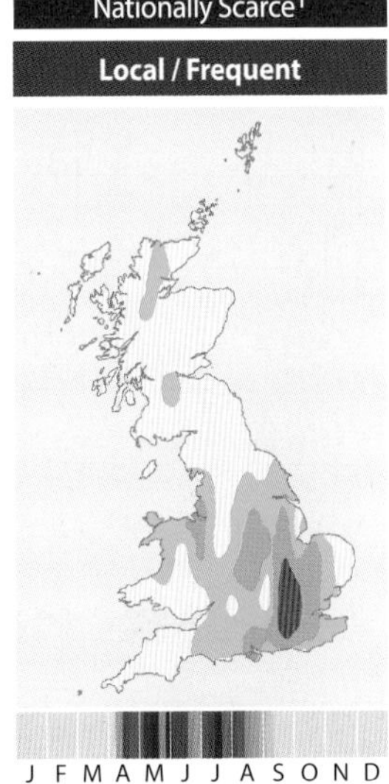

Wing length: 4·5–6 mm

Identification: These are both members of the sub-genus *Neocnemodon*, males of which have a spur on the hind trochanter and females lack dust bars on the face. The five members of this subgenus (*H. pubescens*, *H. vitripennis*, *H. brevidens*, *H. latitarsis* and *H. verrucula*) are very difficult to tell apart, males being best told by microscopic features of the genitalia. Females cannot be identified with any certainty.

Observation tips: *H. vitripennis* is the most frequently encountered member of this rare subgenus and has been captured in numbers in water traps placed in bramble patches; this may indicate its preferred habitat. *H. pubescens* is associated with conifer plantations. Both species are widespread, but most frequent in southern England (the map shows their combined distribution). They do not usually visit flowers.

Heringia species not otherwise covered:

Heringia senilis – Very similar to *H. heringi* (separated on details of the male genitalia) and not regarded as a distinct species by many European authors. Very few, scattered records.

Heringia brevidens [NS]; *Heringia latitarsis* [NS]; *Heringia verrucula* [DD] – These are rare and little known species (placed in the sub-genus *Neocnemodon)*, although the lack of records may be because many recorders avoid them because they are so difficult to identify.

Pipizella

3 British species (2 illustrated)

Small, black hoverflies with rather elongate antennae, and with distinctive wing venation due to the shape of the outer cross-vein. **Accurately determining the identity of individual species is not straightforward, and is not possible in the field or from photos taken in the field. They are best identified by features of the male genitalia.** This is one of several genera which mimic small solitary bees, especially in their behaviour. Larvae feed on root-feeding aphids.

Pipizella viduata / virens

Widespread / Frequent

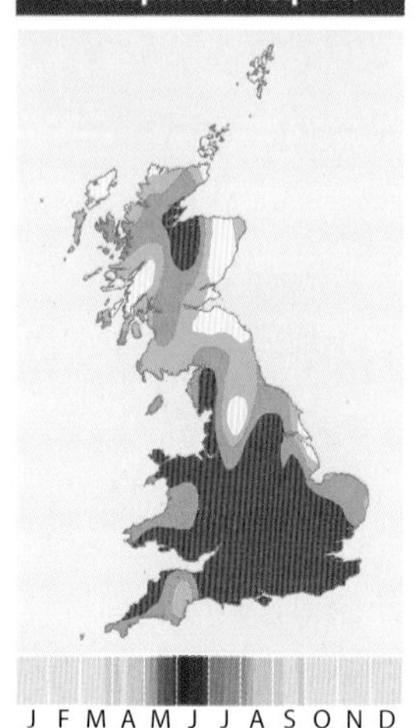

Wing length: 3·75–6·25 mm

Identification: Differentiating these species is best achieved by examining the male genital capsule - which can be readily teased out and twisted round to see the features. Differences in the antennae are also useful: in *P. viduata* the length of the 3rd segment is 2 × the width; in *P. virens* it is around 2½ × the width. Also, the aristae are generally darker in *P. viduata* and yellower in *P. virens* – but this is a less reliable feature.

Similar species: *Heringia* and *Pipiza* species (see table on *page 230*). The Nationally Scarce *Pipizella maculipennis* which can only be reliably identified by microscopic examination of the male genitalia.

Observation tips: Dry grassland and woodland rides where they bask on leaves or bare patches of ground and visit small flowers such as bedstraws. *P. viduata* is widespread, but probably often overlooked; *P. virens* is scarcer. The map shows their combined distribution.

Small solitary bee *Lasioglossum* sp. that *Pipizella*, amongst others, mimics.

Pipizella species not otherwise covered:

Pipizella maculipennis [NS] – Rare species with about 20 records since 1980 scattered across England south of Warwickshire.

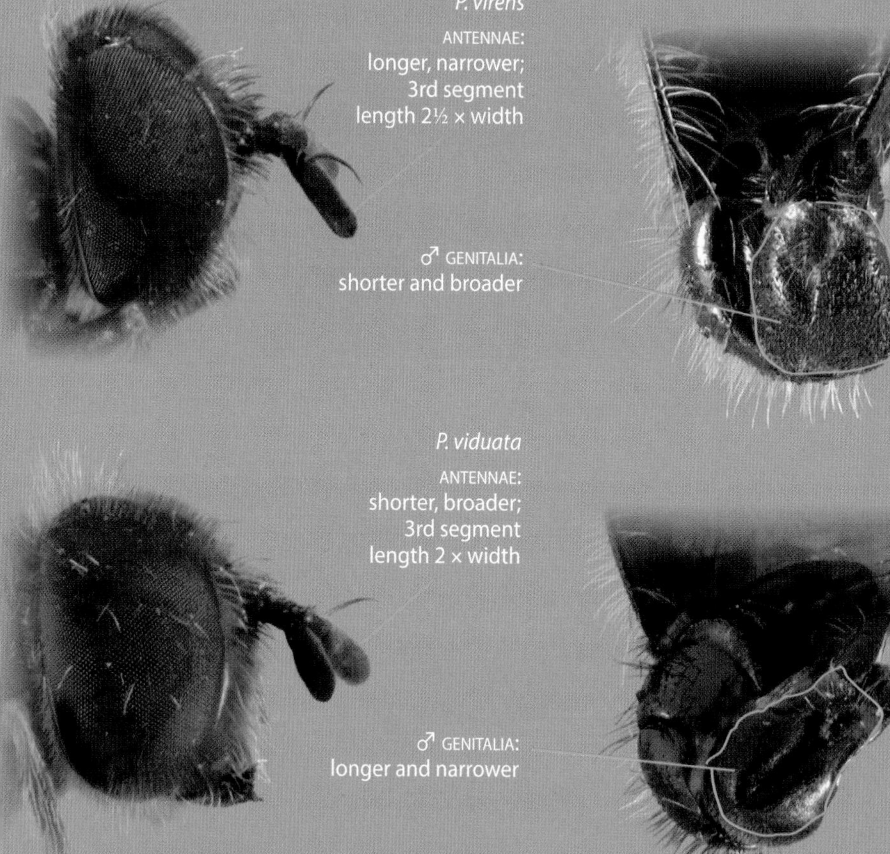

Trichopsomyia — 2 British species (1 illustrated)

Small, black hoverflies resembling *Pipizella* (*p. 236*), but the females have round yellow spots on abdomen segment T2, and both sexes have black hair on the hind tibia (*Pipizella* is entirely black and has pale-haired legs).

Trichopsomyia flavitarsis

Frequent

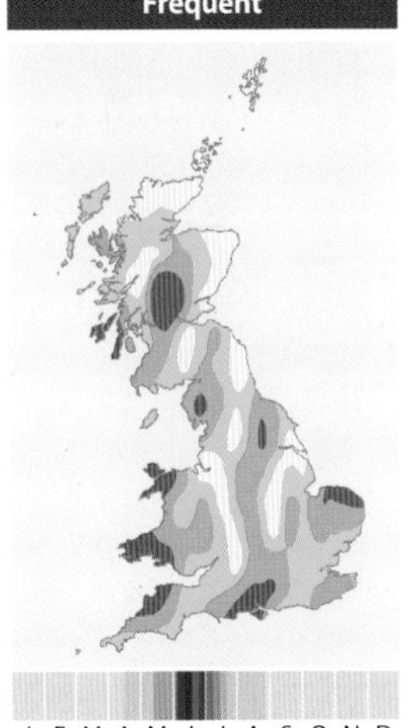

Wing length: 4–6 mm

Identification: The wing venation and general shape are similar to *Pipizella*, but the female abdominal markings are more like *Pipiza*. The long, black hairs on the hind tibia are distinctive, but can only be seen by careful examination under high magnification.

Similar species: *Trichopsomyia lucida* (not illustrated), *Heringia*, *Pipizella* and *Pipiza* species (see table on *page 230*).

Observation tips: Primarily a western and upland species found in damp, rushy pastures, especially where Jointed Rush is abundant. It also occurs in some lowland situations, such as acid heaths and rich fens, and, surprisingly, on some wet calcareous grasslands.

Triglyphus — 1 British species (illustrated)

A tiny black hoverfly resembling *Pipizella* (*p. 236*) which, on close examination, is **immediately identified by its abdomen with only 2 segments visible** (see *page 229*). All other hoverflies have 3 or 4 abdomen segments visible, sometimes more. The larvae feed in aphid galls on Mugwort.

Triglyphus primus

Nationally Scarce

Local

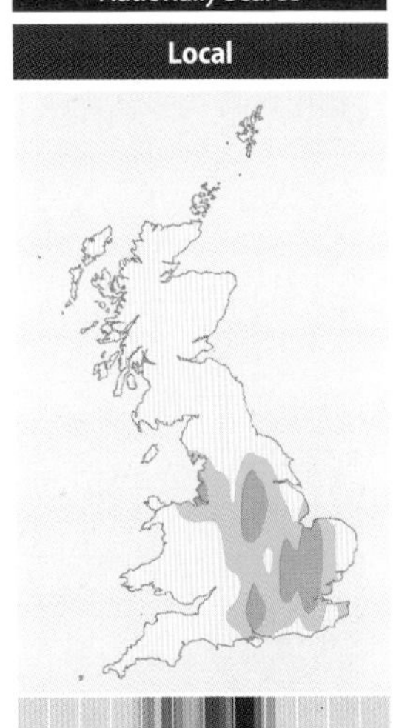

Wing length: 4·25–5 mm

Identification: Often occurs amongst other species within the tribe Pipizini, especially *Pipizella*, and is easily overlooked. The unusual arrangement of the abdomen segments should make it readily recognisable, but this feature does require careful examination under magnification.

Similar species: *Heringia*, *Pipizella* and *Pipiza* species (see table on *page 230*).

Observation tips: Scarce, with a southern distribution. Most records are from localities where Mugwort is abundant (*e.g.* rough grassland, urban waste ground and old industrial sites). Adults tend to visit umbellifers, particularly Wild Parsnip and Wild Carrot.

Trichopsomyia flavitarsis ♀ ×**10**
♀ showing round spots
Trichopsomyia species not otherwise covered:
T. lucida – Discovered in Britain in 2006 in a cemetery in London. It may be confused with yellow-spotted *Pipiza* and could be overlooked.
***Triglyphus*: PIPIZINI**
Triglyphus primus ♀ ×**10**

Guide to Sericomyiini

The single genus within this tribe comprises large, colourful hoverflies with **plumose aristae and the outer cross-veins of the wings not re-entrant**. *Volucella* (*pp. 244–249*) have similar aristae, but the re-entrant outer cross-veins distinguish that genus.

Sericomyia

3 British species (all illustrated)

Large bumblebee and wasp mimics with plumose aristae. Confusion is most likely with the genus *Volucella* (*pp. 244–249*) but *Sericomyia* lacks the re-entrant outer cross-vein of that genus. They can be readily identified from photographs. The larvae are aquatic and live in peaty pools and ditches.

Sericomyia superbiens

Frequent

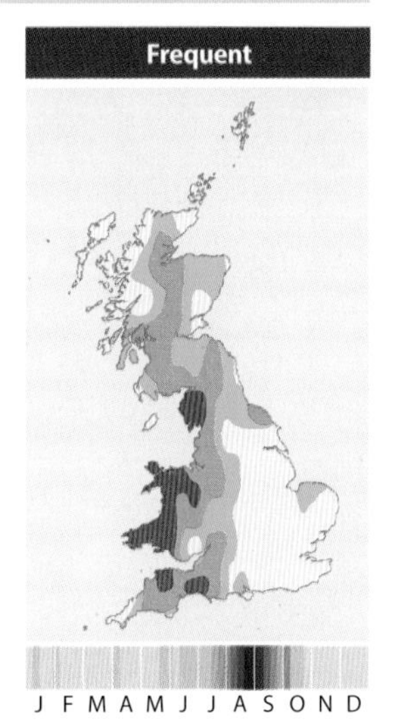

Wing length: 10–13·5 mm

Identification: A large, buff-coloured, bumblebee mimic with dark wing markings. The aristae are strongly plumose and consequently it might be confused with the rare, buff-coloured form of *Volucella bombylans* (*p. 244*), but readily separated by the wing venation which lacks the re-entrant outer cross-vein of that genus. This species was formerly known as *Arctophila superbiens*.

Similar species: The buff form of *Volucella bombylans* is the most likely source of confusion (see above). It may also be confused with the pale form of *Criorhina berberina* and perhaps even *C. floccosa* (both *p. 262*), though the plumose aristae of *Sericomyia* should distinguish it from these two species. Confusion with some forms of *Merodon equestris* (*p. 222* and *p. 31*) has been noted from photographs posted on the internet.

Observation tips: Damp meadows and clearings in the north and west, although it also occurs in Norfolk. Its flight period is unusually late, peaking in late Summer to early Autumn, when it tends to visit blue and purple flowers such as knapweeds and Devil's-bit Scabious.

The plumose aristae of the Sericomyiini.

See *Identifying wasp and bee mimics* on *pages 64–66*.

Sericomyia superbiens ♂ ×**6**

Sericomyia lappona

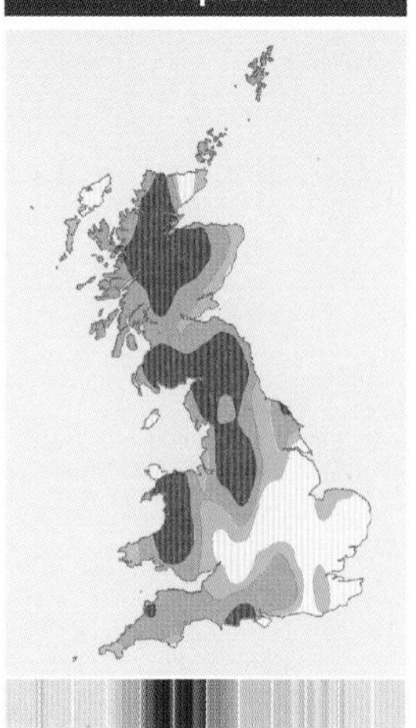

Wing length: 10–13·5 mm

Identification: The narrow, whitish markings on the abdomen and the reddish scutellum combine to make this species easily identifiable.

Similar species: None.

Observation tips: It mainly occurs in May and June, earlier than *S. silentis*, but the flight periods do overlap in the early Summer months. It is primarily a northern and western species but also occurs on southern heathlands. Unlike *S. silentis*, it does not normally move far from breeding locations and is usually seen visiting flowers on or near bogs.

Sericomyia silentis 

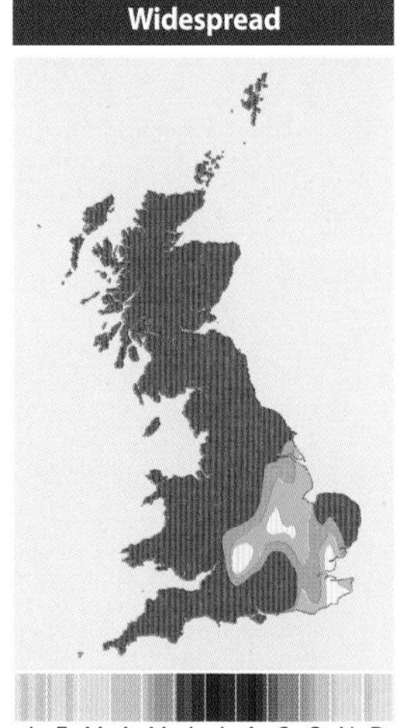

Wing length: 9·5–14·0 mm

Identification: A very large, black-and-yellow wasp mimic that is unlikely to be confused with any other species.

Similar species: Has been misidentified as *Volucella inanis* and *V. zonaria* (both *p. 248*) – the large, black-and-yellow marked members of that genus – but the wing venation is obviously different.

Observation tips: A widespread and abundant upland, northern and western species which favours acid wetlands. In lowland and eastern regions it mainly occurs on heathlands and other acidic habitats. Appears to be very mobile and can be found well away from obvious breeding sites. Adults seem to prefer red and purple flowers such as thistles and knapweeds, but a wide range of flowers may be visited.

See *Identifying wasp and bee mimics* on *pages 64–66*.

Sericomyia lappona ♀ ×**5**

Sericomyia silentis ♂ ×**5**

Guide to Volucellini

This tribe of large hoverflies is represented by a single genus, *Volucella*, which can be easily recognised by the **plumose aristae and wings with the outer cross-vein re-entrant**. Members of the Sericomyiini (*pp. 240–243*) have similar aristae, but the outer cross-vein is not re-entrant.

Volucella

5 British species (all illustrated)

Amongst the largest British hoverflies and instantly recognisable. The five British species are quite diverse in general appearance and are fairly easy to separate. The larvae of all except *V. inflata* live in the nests of social wasps or bumblebees.

Volucella bombylans

Widespread

Wing length: 8–14 mm

Identification: A polymorphic bumblebee mimic. The plumose aristae separate it from most other bumblebee mimics apart from *Sericomyia superbiens,* which does not have a re-entrant outer cross-vein. Two forms are common: a black form with a red-orange 'tail' resembling *Bombus lapidarius*, and a black-and-yellow form which resembles *B. lucorum* (see insets *opposite*). A third, entirely buff-coloured form, resembling *B. pascuorum* (see *page 263*), is much scarcer.

Similar species: Other bumblebee mimics such as *Sericomyia superbiens* (*p. 240*), *Criorhina berberina*, *C. floccosa*, *C. ranunculi* (*pp. 260–263*) and *Pocota personata* (*p. 264*).

Observation tips: A widespread and common visitor to a wide range of flowers. Males are highly territorial, defending a sunlit leaf and foraying out to investigate possible mates. The larvae live in bumblebee nests where they are scavengers amongst debris in the bottom of the nest.

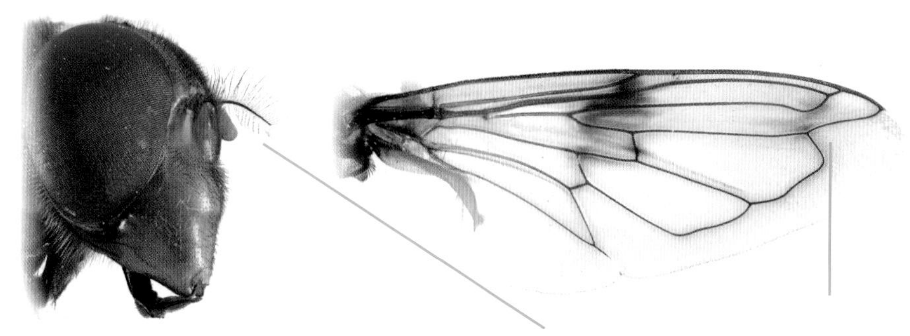

Volucella: plumose aristae (LEFT) and wing showing the distinct re-entrant outer cross-vein (RIGHT).

See *Identifying wasp and bee mimics* on *pages 64–66*.

The two common forms of *Volucella bombylans* resemble different bumblebees: the black form resembles *Bombus lapidarius* [TOP INSET] and the black-and-yellow form of *Bombus lucorum* agg. [BOTTOM INSET]

Bombus lapidarius

Volucella bombylans ♀ black form ▲ ♀ black-and-yellow form ▼ ×**6**

Bombus lucorum agg.

Volucella inflata

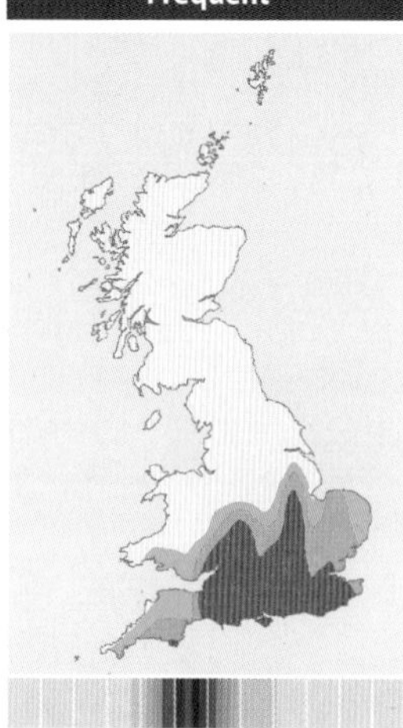

Wing length: 11–12·75 mm

Identification: A medium-sized almost globular hoverfly with golden-orange markings on abdomen segment T2 and strong black wing clouds. It can only be confused with *V. pellucens*, which has similar shaped, but white, abdominal markings. The markings of *V. pellucens* can sometimes be a little creamy-yellowish, occasionally leading to confusion, but they are never anything like the golden colour of *V. inflata*.

Similar species: *Volucella pellucens*.

Observation tips: A woodland species which is locally abundant in southern England and South Wales. The adults are regular flower visitors, favouring Bramble and shrubs such as Dogwood and Wild Privet. Males fly rapidly around and through such bushes seeking mates. Unlike other British *Volucella*, the larvae inhabit sap runs.

Volucella pellucens

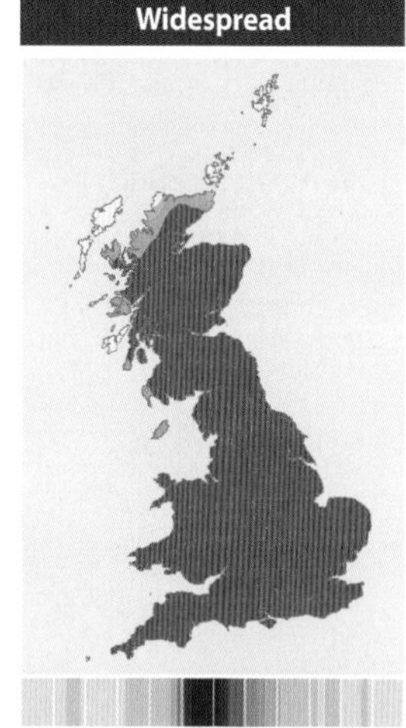

Wing length: 10–15·5 mm

Identification: One of the most obvious British hoverflies, with large white markings on the abdomen segment T2 contrasting with the otherwise black thorax and abdomen. It is commonly known as the Great Pied Hoverfly.

Similar species: *Volucella inflata* is similar. *Leucozona lucorum* (*p. 112*) and *Cheilosia illustrata* (*p. 162*) have vaguely similar colour patterns but lack the distinctive wing venation of *Volucella*.

Observation tips: Widespread and abundant. Adults are generally found in sheltered situations such as woodland rides and tree-lined paths. Males hover at around head height and defend a beam of sunshine. They dart off to investigate intruders or possible females and then return to hover in the same spot. Both sexes visit a wide range of flowers and are often seen in gardens. Larvae live in the nests of a range of social wasps, where they are scavengers amongst the debris in the bottom of the nest cavity.

See *Identifying wasp and bee mimics* on *pages 64–66*.

Volucella inflata ♀ ×**4**

♂ – *Volucella pellucens* ×**4** – ♀

Volucella inanis

Frequent

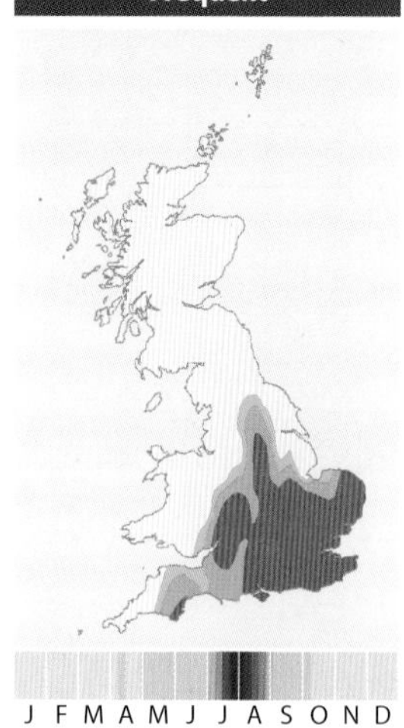

Wing length: 12·25–14·25 mm

Identification: A large yellow-and-black wasp mimic which could only possibly be confused with *V. zonaria.*

Similar species: *Volucella zonaria* and possibly *Sericomyia silentis* (*p. 242*), though the difference in wing venation should be easy to see.

Observation tips: Now a widely distributed species in southern England, north to Yorkshire. It has undergone a dramatic expansion in range since about 1995 having previously been confined to southern England, especially the London area. Larvae live in the nests of ground-nesting social wasps where they feed on the wasp grubs. Since wasps often build their nests in buildings, this hoverfly frequently turns up indoors, often being seen on windows. Adults visit a wide range of flowers in mid- to late Summer.

Volucella zonaria

Frequent

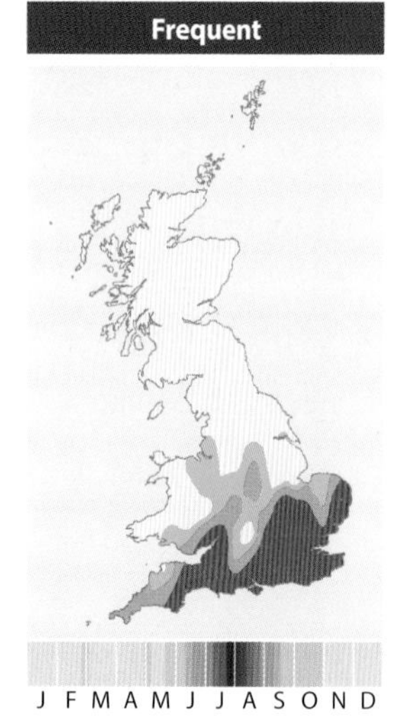

Wing length: 15·5–19·5 mm

Identification: The largest British hoverfly and often described as a Hornet mimic. It should not be mistaken for any hoverfly other than *V. inanis* (see table opposite).

Similar species: *Volucella inanis.* Reports of *V. zonaria* from northern England and southern Scotland have proved to be *Sericomyia silentis* (*p. 242*).

Observation tips: Widespread and increasingly abundant in southern England south of Cheshire and Humberside. Usually seen visiting garden flowers such as *Buddleja.* A relatively recent colonist that arrived on the south coast of England in the late 1930s. It became established, mainly in the London area, resulting in a steady stream of notes in the entomological press from the 1940s through to the 1970s. Consequently, its distribution changes are amongst the best documented of any hoverfly. Since about 1995, it has expanded its range rapidly, and this trend is continuing. Larvae live in the nests of social wasps that build in tree cavities, including the Hornet. They are scavengers amongst the debris in the bottom of the nest cavity.

See *Identifying wasp and bee mimics* on *pages 64–66.*

V. inanis: ABDOMEN: T2 **yellow**; **underside largely yellow**; SCUTELLUM: dull yellowish-grey
V. zonaria: ABDOMEN: T2 **chestnut**; **underside with broad black bars**; SCUTELLUM: usually chestnut

Guide to Xylotini

This is a heterogeneous tribe composed of 10 genera, seven of which contain just a single species. The humeri are hairy; a feature which can be easily seen because the head and the front of the thorax are well separated. Many of the genera are instantly recognisable - with practice. Some have distinctly enlarged and ornamented hind femora; others are bumblebee mimics; and others still are rather elongate and resemble ichneumon wasps or sawflies of the genus *Macrophya*. **The relative position of the vein R-M in relation to the discal cell is diagnostic: it joins the upper margin of the discal cell at a point at or beyond the middle of this cell.**

1

a) Distinctive species:

Metallic greenish-black hoverfly with orange-brown wings and orange legs
Caliprobola speciosa p. 254

Abdomen black and orange
Blera fallax p. 252

Both *Blera* and *Caliprobola* are rare.

b) Narrow-bodied species with grey dusted thorax side:

Hind femora greatly enlarged, without flange
Syritta pipiens p. 266

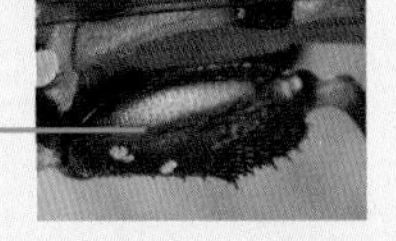

Hind femora greatly enlarged, with flange
Tropidia scita p. 266

c) Bumblebee mimics ▶ 2

d) Honey-bee mimics ▶ 3

e) Sawfly mimics ▶ 4

BUMBLEBEE MIMICS

2 *from* **1c**

a) Distinctive face with elongated, downward-pointing mouth edge

Criorhina *pp. 260–263*

b) Head small in comparison to body

Pocota personata *p. 264*

3 *from* **1d**

HONEY-BEE MIMICS

a) Hind femora not enlarged:

Distinctive face-shape

Criorhina *pp. 260–263*

Criorhina asilica

b) Hind femora enlarged:

Underside of thorax bare

Brachypalpus laphriformis *p. 256*
– pictured right

Underside of thorax hairy

Chalcosyrphus eunotus *p. 258*

4 *from* **1e**

SAWFLY MIMICS

a) Legs partly yellow

Xylota *pp. 268–271*

Chalcosyrphus nemorum *p. 258*
See account for further information on how to distinguish *Chalcosyrphus nemorum* from some extremely similar *Xylota*.

Xylota segnis

b) Legs black

Brachypalpoides lentus *p. 254*

Brachypalpoides lentus

Blera

1 British species (illustrated)

This single species in this genus is very distinctive and extremely rare. The larvae inhabit rotting cavities in pine stumps. Its natural habitat is probably shattered stumps of windblown pines that had been weakened by the fungus *Phaeolus schweinitzi*. Recent studies have shown that suitable habitat can be created by making holes in pine stumps with a chainsaw.

Blera fallax

Wing length: 8–9·5 mm

Identification: A very distinctive hoverfly with a black thorax and base to the abdomen that contrasts with the orange-red terminal segments. This combination of colours is unique amongst British hoverflies.

Similar species: None.

Observation tips: Only two small populations are known, both in stumps in pine plantations along the River Spey in central Scotland. It probably takes larvae several years to develop and as the population is very small adults are rarely seen.

Typical Xylotini wing (*Xylota*).

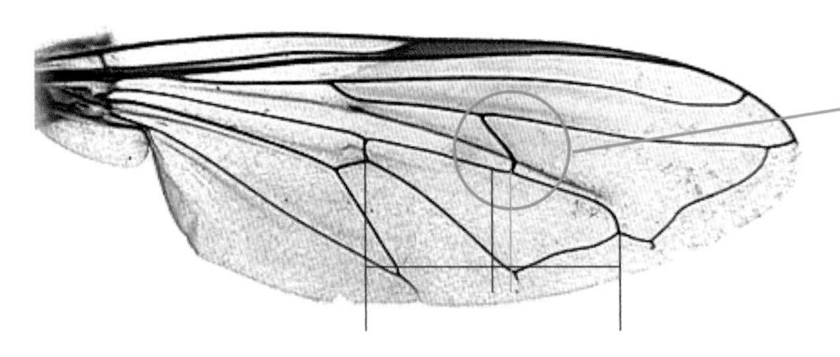

The Xylotini are enormously varied in form but a consistent character is the position of the R-M vein that reaches the upper margin of the discal cell at the middle (*Syritta*) or a point beyond the middle of this cell.

Blera fallax ♂ ▲ ♀ ▼ ×5

Brachypalpoides

1 British species (illustrated)

A very distinctive hoverfly which is a denizen of old woodlands, especially oakwoods. It is closely related to *Xylota* (*pp. 268–271*) and the adults behave in a similar way – running over leaves as they feed on honeydew and the pollen grains trapped in this sticky covering. Adults may be mimicking sawflies such as *Tenthredo* and *Macrophya* and/or spider hunting wasps. Larvae live in the soft, wet, rotten wood of large, dead tree roots.

Brachypalpoides lentus

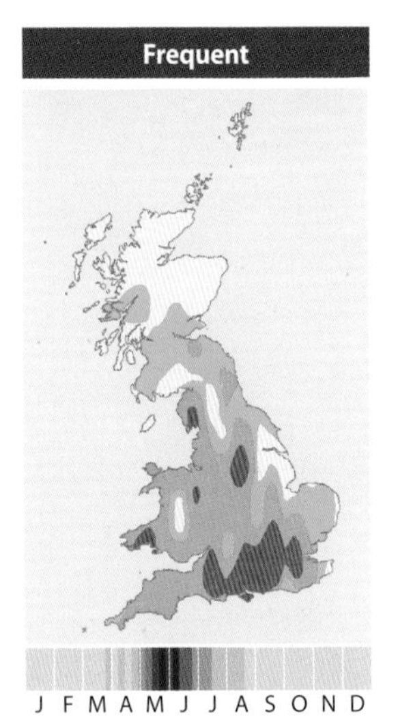

Wing length: 10–12 mm

Identification: The combination of black body and legs, together with deep red markings on the base of the abdomen, are unique.

Similar species: Confusion is only likely with *Xylota segnis* (*p. 270*) and *X. tarda* (not illustrated), which have partly yellow legs and orange, rather than blood-red, abdominal markings.

Observation tips: Females can often be found flying low down around the roots of oaks and other large deciduous trees. Otherwise, adults are usually observed as they feed on the surface of sunlit leaves in woodland rides and clearings.

Caliprobola

1 British species (illustrated)

This is one of the largest and most spectacular of Britain's hoverflies. The larvae live in soft, wet, decaying wood with a porridge-like consistency in the dead roots of old Beech trees.

Caliprobola speciosa 

Near Threatened

Rare

Wing length: 11–12·5 mm

Identification: Readily identified by its unusual metallic greenish-black colour, orange legs and relatively long, narrow, orange-brown tinged wings.

Similar species: No other species found in the UK resembles *C. speciosa*.

Observation tips: Confined to the New Forest and Windsor Great Park, where it can occasionally be abundant in suitable areas. Although a strong flier, this species has not shown any sign of dispersing from its two known localities. This makes it potentially very vulnerable to any change, such as drought, which might lead to the loss of ancient Beech trees.

Brachypalpoides lentus ♂ ×**5**

Models that this species and others with a similar abdominal pattern, such as *Xylota segnis* (*p. 270*) may mimic: a sawfly *Macrophya annulata* [TOP INSET]; a spider-hunting wasp *Priocnemis perturbator* [BOTTOM INSET].

Caliprobola speciosa ♀ ×**5**

Caliprobola: XYLOTINI

Brachypalpus

1 British species (illustrated)

A woodland species that is quite a convincing bee mimic. It is most frequently likened to an *Osmia* bee as males often fly around dead timber in a similar manner to male *Osmia*. The larvae are found in wet rot holes in large, old, deciduous trees, especially oaks.

Brachypalpus laphriformis

Local

Wing length: 8·5–10·75 mm

Identification: This bee mimic has inflated hind femora, which are arched in males. The most likely confusion is with *Chalcosyrphus eunotus* (*p.* 258), the males of which also have inflated hind femora, though these are straight. However, the underside of the thorax, between the middle and hind legs, is bare (hairy in *C. eunotus*). Other similar species include *Criorhina asilica* (*p.* 260), which does not have inflated hind femora; it also has diffuse abdominal markings.

Similar species: *Chalcosyrphus eunotus* and *Criorhina asilica* (see above).

Observation tips: Occurs in the vicinity of old trees with wet rot holes, especially oaks. It is largely a western species, occurring as far north as the Lake District.

Osmia bee – in flight, *Brachypalpus* mimics species from this genus.

See *Identifying wasp and bee mimics* on *pages 64–66.*

Brachypalpus laphriformis ♂ ×**7**

hind femora

Chalcosyrphus – **straight**

Brachypalpus – **rounded**

underside of thorax between the middle and hind legs

Chalcosyrphus – **hairy**

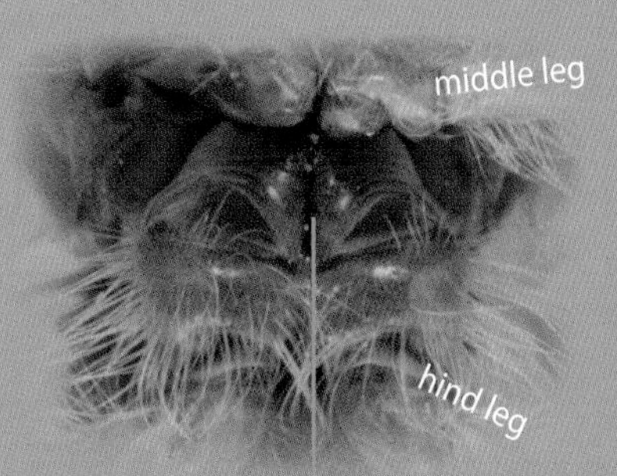

Brachypalpus – **bare**

Chalcosyrphus

2 British species (both illustrated)

This is a very variable genus in Europe and the two British species are rather different from one another. **Unlike related genera, such as *Xylota* (*pp. 268–271*), the underside of the thorax, between the middle and hind coxae, is hairy.** Larvae live in decaying sap under the bark of waterlogged timber and, as a consequence, adults are generally found in or near wet woodland.

Chalcosyrphus eunotus

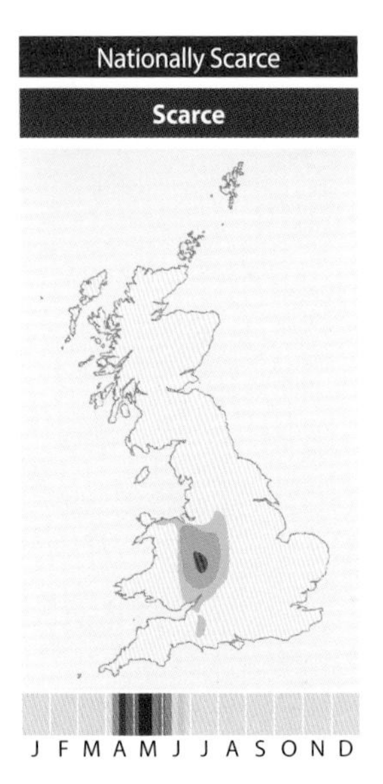

Wing length: 9·5–10·5 mm

Identification: This is a small honey-bee mimic with straight, inflated hind femora and a hairy underside to the thorax, between the middle and hind legs (bare in *B. laphriformis*).

Similar species: Most likely to be confused with *Brachypalpus laphriformis* (see *page 256* for differences). *Criorhina asilica* (*p. 260*) and older, paler examples of *C. floccosa* (*p. 262*) may also cause confusion but both these species lack inflated hind femora and have a distinctive face-shape.

Observation tips: This is a highly localised species of 'dingle woodlands' *i.e.* deep, wooded, stream gorges. The larvae live under the bark of partially submerged timber and this species is a flagship for a suite of invertebrates that are closely associated with log jams. This has become a rare habitat because log jams tend to get removed from streams and rivers by land managers to reduce the risk of flooding.

Chalcosyrphus nemorum 

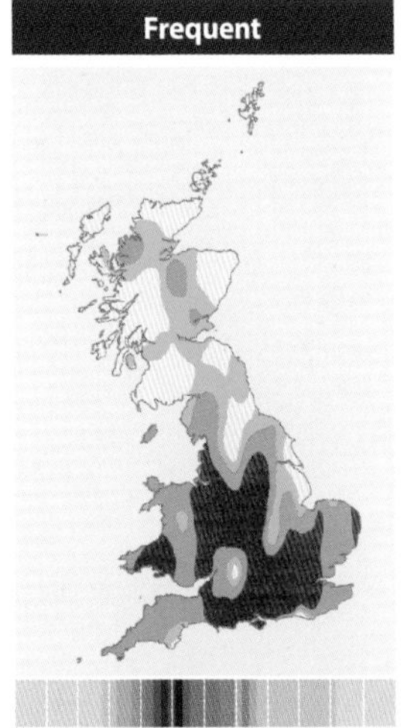

Wing length: 6·5–8·25 mm

Identification: Closely resembles *Xylota abiens* in appearance but with the hind femora expanded towards the apex (see image *opposite*). The abdomen segments of *C. nemorum* tend to be wider than long, making the body look less elongate than *X. abiens*. Both these characters are tricky to assess unless specimens are available for comparison. However, *C. nemorum* has a hairy underside to the thorax which readily distinguishes it from other similar-looking species.

Similar species: *Xylota abiens* and the smaller *X. jakutorum* (*p. 268*) are most likely to cause confusion.

Observation tips: A species of wet woodlands, found particularly in Alder carr. Adults are regular visitors to flowers, especially buttercups.

See *Identifying wasp and bee mimics* on *pages 64–66*.

Chalcosyrphus eunotus ×**7**

The hairy underside of the thorax distinguishes *Chalcosyrphus* from other Xylotines

hind femora

Chalcosyrphus nemorum
expanded towards apex

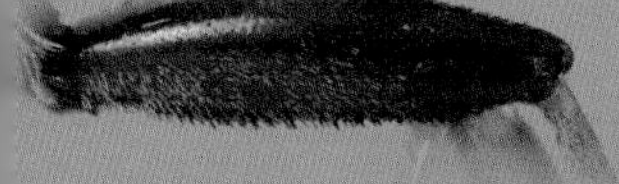

Xylota abiens
not expanded towards apex

Chalcosyrphus nemorum ♀ ×**7**

Criorhina

4 British species (all illustrated)

Medium sized to large bumblebee or honey-bee mimics with a **distinctive face-shape in which the mouth edge is elongated downwards**. The larvae live in wet, decaying subterranean timber. Females are often seen flying around the bases of tree stumps and large trees looking for egg-laying sites, and 'bees' behaving like this are therefore worth investigating.

Criorhina asilica

Frequent

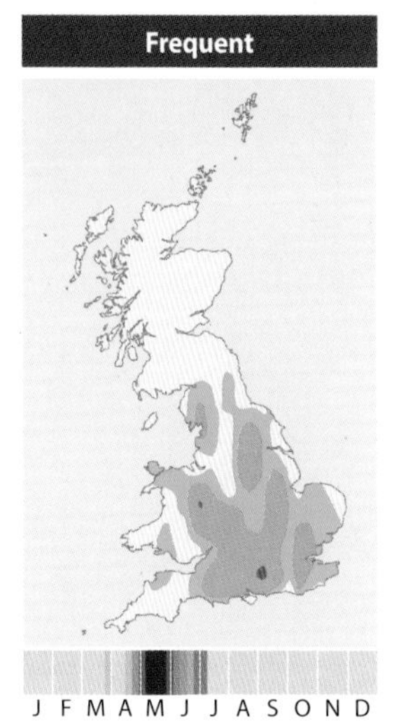

Wing length: 9·5–11·25 mm

Identification: A good honey-bee mimic – although it has also been suggested that the behaviour of the male mimics solitary bees, especially *Andrena carantonica*. There are few species with which it can be confused, although it remains tricky to identify as there are few definite features that make it immediately distinct and it is sexually dimorphic. The greyish-orange markings on abdomen segments T2 and T3 are useful features.

Similar species: *Brachypalpus laphriformis* (*p. 256*) and *Chalcosyrphus eunotus* (*p. 258*), both of which have enlarged hind femora. Confusion with *Criorhina floccosa* and *C. berberina* (both *p. 262*) may be possible in poor light.

Observation tips: A woodland species which occurs widely in England and Wales. Males fly low and rapidly over low-growing vegetation in a manner similar to males of the solitary bee *Andrena carantonica*. Both sexes are regular visitors to flowers such as Hawthorn and Dogwood. It seems to have become more frequent in recent years.

Criorhina ranunculi

Frequent

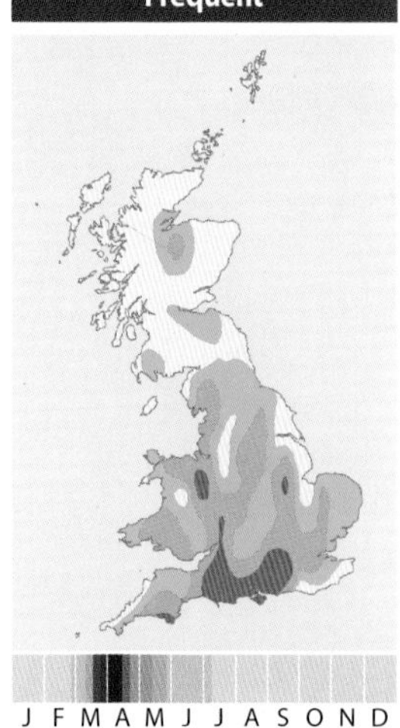

Wing length: 11·25–14 mm

Identification: Bigger and darker than the other members of the genus. The expanded, arched hind femora make it very distinct and unlikely to be confused with other bumblebee mimics. There are several colour forms with red or white 'tails' that mimic different bumblebee species.

Similar species: Bumblebee mimics such as *Eristalis intricaria* (*p. 206*), *Merodon equestris* (*p. 222*) and *Volucella bombylans* (*p. 244*); none of these have the distinctive *Criorhina* face-shape.

Observation tips: An early Spring species which visits the flowers of Blackthorn, willows and Wild Cherry, often staying high up. Females can be found prospecting around the roots of trees, including birches on heathland and oaks in woodlands. Mainly a southern species, although there are scattered records north to the Moray Firth.

See *Identifying wasp and bee mimics* on *pages 64–66*.

Criorhina asilica ♀ ×**4**

♂ – *Criorhina ranunculi* ×**4** – ♀

Criorhina berberina

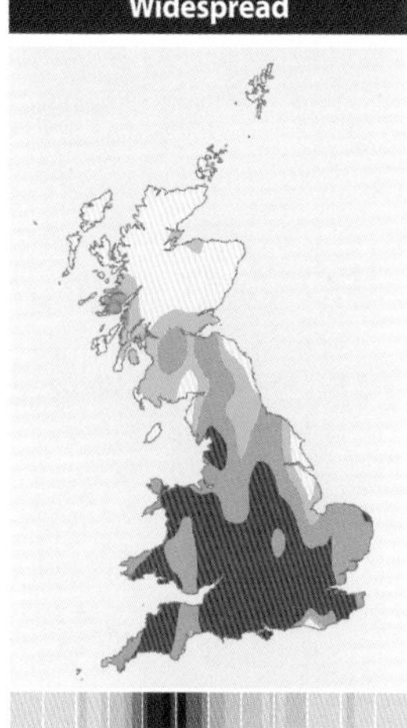

Wing length: 8–12 mm

Identification: A bumblebee mimic with two distinct colour forms. The typical form is dark with a buff stripe across the front of the thorax and a pale-haired 'tail'. The other form, var. *oxyacanthae*, entirely buff-coloured, mimics the carder bee *Bombus pascuorum* and resembles *Sericomyia superbiens* and the buff form of *Volucella bombylans*.

Similar species: Other buff-coloured bumblebee mimics such as *Sericomyia superbiens* (*p. 240*) and the buff form of *Volucella bombylans* (*p. 244*) (both of which have plumose antennae and dark markings on the wings) could be confused with var. *oxyacanthae*. Confusion is also possible with *Criorhina floccosa*, which has warmer-toned orange fur and distinct tufts of pale hair at the sides of the abdomen near the base.

Observation tips: This is a woodland and hedgerow species which occurs widely across mainland Britain. The adults are regular visitors to a wide range of flowers including Hogweed.

Criorhina floccosa

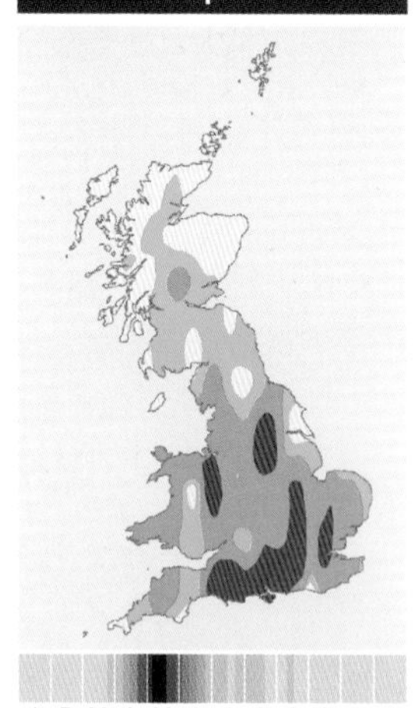

Wing length: 10–13 mm

Identification: A mimic of bumblebees such as the carder bee *Bombus pascuorum*. The most distinctive feature is the tufts of pale hair at the sides of the abdomen near the base. It is not always straightforward to separate from *C. berberina* var. *oxyacanthae* if these hair tufts are not obvious, but it tends to have more orange-brown fur.

Similar species: Other buff-coloured bumblebee mimics such as *Criorhina berberina* var. *oxyacanthae*, *Sericomyia superbiens* (*p. 240*) and the buff form of *Volucella bombylans* (*p. 244*).

Observation tips: This is a woodland species which frequently basks on sunlit leaves and visits a range of flowers such as Hawthorn and Dogwood. It often shows interest in the bases of a range of deciduous trees and the hollow trunks of ancient Ash pollards.

See *Identifying wasp and bee mimics* on *pages 64–66*.

Criorhina berberina ♀ ×**5**

The typical form (LEFT) is very different from var. *oxyacanthae* (BELOW).

This form of *C. berberina* and *C. floccosa* both mimic the carder bee *Bombus pascuorum* (INSET)

Bombus pascuorum

Tufts of pale hair at the base of the abdomen

Criorhina floccosa ♂ ×**5**

Pocota

1 British species (illustrated)

A very convincing bumblebee mimic, the adults even emitting a loud buzz that sounds like an angry bee and, when disturbed, lifting a hind leg in the same manner as a bumblebee. **It is distinct from other hoverflies because it has a disproportionately small head in comparison to the body**. The larvae live in wet rot holes in Beech and it has been suggested that they prefer holes that are high up.

Pocota personata

Nationally Scarce

Scarce

Wing length: 11–13 mm

Identification: Should not be confused with other bumblebee mimics because of the small head. The yellow hairs on the thorax and abdomen are strikingly lemon-yellow, adding to its unique appearance.

Similar species: In the absence of careful inspection, may be misidentified as *Criorhina berberina* (*p. 262*) or *Volucella bombylans* (*p. 244*).

Observation tips: There are a few scattered records in England and Wales north to County Durham. However, it may not be as rare as records suggest, as the very short flight period (peaking late May to early June) means it can be easily overlooked. Searching for larvae in rot holes may be a more efficient way of finding this species. Both sexes occasionally visit flowers.

A mating pair of *P. personata*.

See *Identifying wasp and bee mimics* on *pages 64–66*.

Pocota personata ♂ ×**5**

Syritta

1 British species (illustrated)

A small, dark, narrowly-built hoverfly with greatly enlarged hind femora and the sides of the thorax heavily dusted ash-grey. Males are highly territorial; opponents will face up, forcing one another forwards and backwards until one gives up. Larvae live in wet, decaying matter such as compost, manure and silage, but not actually in water.

Syritta pipiens

Widespread

Wing length: 4·25–7 mm

Identification: This species is unmistakable, provided you get close enough to see the characters described above.

Similar species: There should be no confusion with other British species, although there is a very similar species, *Syritta flaviventris*, in southern Europe.

Observation tips: Widely distributed and abundant, and commonly seen in gardens; only absent from more exposed and upland situations. Adults visit a wide range of flowers. Occasionally, there are mass occurrences that may be a result of immigration. Tends to be abundant later in the season and into the Autumn, and can be numerous on Ivy flowers.

Tropidia

1 British species (illustrated)

A medium-sized, elongate hoverfly with conspicuous orange markings on the abdomen. It is immediately recognisable when examined closely – the hind femora have a pronounced triangular 'tooth'. Found in reedbeds, the larvae live in wet, decaying vegetation around reed bases.

Tropidia scita

Frequent

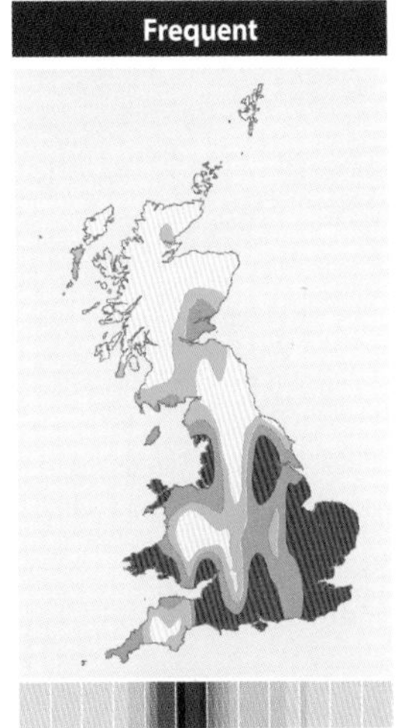

Wing length: 5·5–8·25 mm

Identification: The pronounced triangular 'tooth' on the hind femur is distinctive – only *Merodon equestris* (*p. 222*), a bumblebee mimic and hence very different looking, has anything similar.

Similar species: *Xylota segnis* (*p. 270*) and *X. tarda* (not illustrated – see *page 268*) are a somewhat similar shape and colour but lack the femoral 'tooth'.

Observation tips: Often abundant in beds of Common Reed, Bulrush and other tall emergent vegetation in ditches, ponds and fens. Adults can be found flitting amongst reed stems and visiting flowers, but not usually very far from breeding sites. It has been found widely around the coast as far north as the Outer Hebrides, but only seems to occur inland in the south, especially in East Anglia and the south-east.

Syritta: XYLOTINI

Syritta pipiens ♂ × **10**

Tropidia: XYLOTINI

Tropidia scita ♂ × **10**

Xylota

7 British species (3 illustrated)

Rather elongate hoverflies that generally mimic sawflies or spider-hunting wasps (see *page 255*). Often seen running over leaves feeding on a mixture of pollen and honeydew rather than visiting flowers. The larvae live in decaying timber or wood debris.

Xylota jakutorum

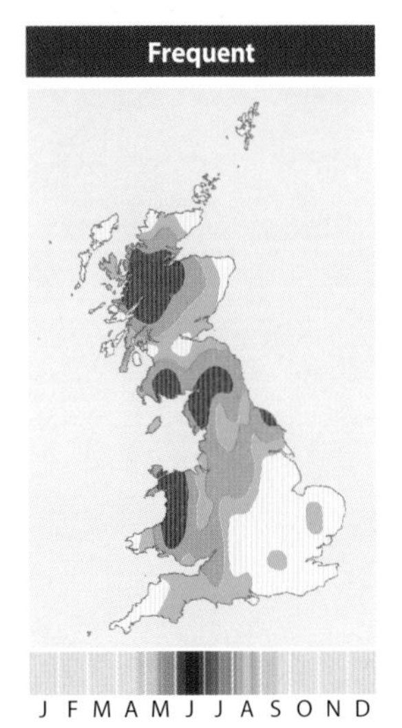

Wing length: 7·5–8 mm

Identification: This species, with orange-red spots on the abdomen, is most likely to be confused with *X. florum*. Male *X. florum* have a long, narrow abdomen due to the abdomen segments being elongated. In comparison, *X. jakutorum* has wider segments and, consequently, a shorter abdomen overall. It also has more extensive and conspicuous spines on the underside of the hind femur than *X. florum*.

Similar species: The Nationally Scarce *Xylota abiens* (not illustrated, but see *page 259*) and *X. florum* (see above).

Observation tips: Northern and western Britain. In the 19th Century it was regarded as a Caledonian pine wood species, but has spread southwards in pine plantations and may still be spreading south-eastwards. Closely associated with decaying conifer timber, especially in recently felled areas, where it breeds in the rotting stumps. It can be abundant at buttercups and other flowers in rides and clearings in conifer plantations.

Xylota species not otherwise covered:

X. abiens [NS]	– Most similar to *Chalcosyrphus nemorum* (see *page 258* for differences). A scarce species of southern England, but with scattered records north to Cumbria.
X. florum	– Most similar to *X. jakutorum* (differences discussed above). Widespread in England and Wales, but less frequent than *X. jakutorum* and generally found in damp woodland.
X. tarda [NS]	– Difficult to separate from *X. segnis* (*p. 270*). A rather rare but widespread species associated with Aspen and other poplars.
X. xanthocnema [NS]	– Most similar to *X. sylvarum* (see *page 270* for differences). A scarce species of Wales and the southern half of England. Usually in or near ancient broadleaved woodland.

Xylota jakutorum ♂ ×**7**

Hind femora

X. jakutorum

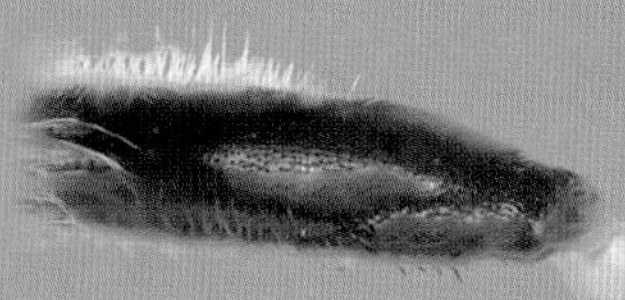

X. florum

Xylota segnis

Wing length: 7–9·5 mm

Identification: The combination of the elongate body shape, orange-red markings on the abdomen and partly yellow legs makes this species distinct, with few possibilities for confusion.

Similar species: The most likely confusion species is the Nationally Scarce *Xylota tarda* (not illustrated), which is rather rare and associated with Aspen. This species has the orange marking on the abdomen divided by a black band on the end of segment T2 , but since *X. segnis* can sometimes show a trace of this feature, microscopic examination of the spines on the hind leg is required for accurate identification. At first glance, *Tropidia scita* (*p. 266*) or *Platycheirus granditarsus* (*p. 88*) are possible confusion species.

Observation tips: Widespread and abundant.
Easily recognised by its behaviour: it scuttles back and forth, sweeping from side to side across leaves, collecting pollen and honeydew. It is rarely found at flowers, at least in the south. In Scotland, however, it can be found visiting flowers such as Meadowsweet and Angelica.

Xylota sylvarum

Wing length: 7–12mm

Identification: Two species of *Xylota* have a golden 'tail' – *i.e.* a patch of golden hairs at the end of the abdomen (*X. sylvarum* and the Nationally Scarce *X. xanthocnema*). *X. sylvarum* is the bigger and bulkier of the two and can be separated from *X. xanthocnema* by the colouration of the hind tibiae (see *opposite*). These are partially black in *X. sylvarum* but at most reddish-orange from some angles in the more dainty *X. xanthocnema*.

Similar species: Only *Xylota xanthocnema* (see above).

Observation tips: Widespread in England and Wales but scarcer in Scotland (*X. xanthocnema* is a scarce southern species). Found in deciduous woodland and a regular flower visitor, especially to white umbellifers such as Hemlock Water-dropwort. Also feeds on leaf surfaces, like *X. segnis*.

♂ *X. segnis* has two rows of spines beneath the hind femur and a trochanter with long spurs

Xylota segnis ♂ ×**7**

Xylota sylvarum ♀ ×**7**

X. xanthocnema has yellow hind tibia

X. sylvarum has apical half of hind tibia darkened

Microdon

4 British species (4 illustrated)

Dumpy hoverflies with short, pointed wings and long, forward-pointing antennae. The larvae live in ants' nests where they feed on the eggs and brood of the ants. Adults are not strong fliers and tend to be found close to breeding sites. They are hard to find and all species are most easily located by looking for larvae and puparia in ants' nests.

Microdon analis

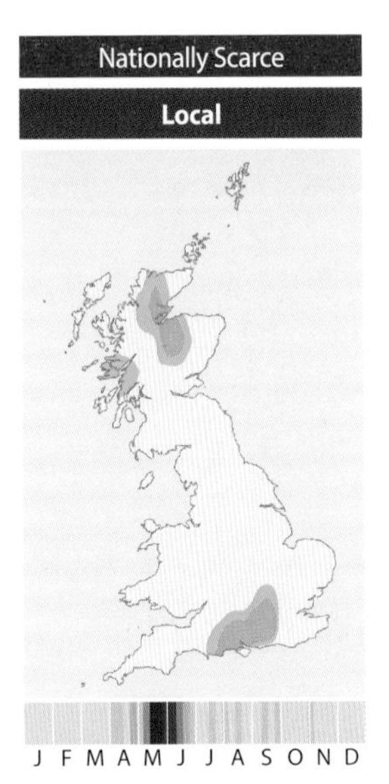

Wing length: 6·75–8·25 mm

Identification: This species has a dark scutellum and the hairs on the disc of the thorax are pale. It might be confused with *M. devius* (*p. 274*) except that the latter has a zone of partly black hairs on the thorax between the wing bases.

Similar species: *Microdon* are unlike other hoverflies, but the species are superficially similar and require careful examination.

Observation tips: Larvae live in the nests of black ants of the *Lasius niger/platythorax* complex on heathlands. Adults rarely fly far and do not appear to be flower visitors. The distribution is disjunct: they occur on the heaths of central southern England and in Caledonian pine woods in the north of Scotland.

Microdon - armoured intruders

The larvae feed on ant larvae and pupae within ants' nests, and are superbly adapted to this life. They are hemispherical, tank-like animals with a distinctly sculptured upper surface and a dense fringe of hairs that help to prevent attack by ants. The head is usually hidden underneath the larva and only protrudes through the fringe of protective hairs to grab prey. The shape of the larva allows it to cling tightly to surfaces, thus making it difficult for ants to penetrate the defences. The larvae absorb the ants' pheromones; making them chemically indistinguishable from the ants themselves. This chemical protection is passed on to the next generation via the egg and allows the 1st instar larvae to enter a host nest.

The relationship between ants and *Microdon* colonisation is complex and is not the same in all species. Recent studies of the biology of *Microdon mutabilis* and *M. myrmicae* has highlighted some differences. In *M. mutabilis*, the farther a female was moved from the original ant colony from which she emerged, the less likely were her offspring to be able to enter an ants' nest. However, *M. myrmicae* did not seem to show this relationship. These biochemical interactions between *Microdon* larvae and ants opens an intriguing window onto the issues that may affect the conservation of our rarer *Microdon* species.

So far, *M. mutabilis* and *M. myrmicae* can only be distinguished by features of their larvae and puparia. It is possible that there are also two or more species within *M. analis* that may similarly be distinguishable only by features of the immature stages.

Collection of *M. analis* puparia under bark

M. analis puparium with ant

Key differences between British *Microdon* species.

SPECIES	SCUTELLUM	THORAX HAIRS
M. analis	**dark**	hairs on thorax disc **pale**
M. devius		hairs on thorax between wing bases **partly black**
M. myrmicae	**red-brown**	adults cannot be separated; they are distinguished by features of mature larvae/puparia.
M. mutabilis		

Microdon devius 

Wing length: 6·25–9·25 mm

Identification: The largest of the British *Microdon* species. Separated from the others by the dark scutellum and the presence of a zone of partly black hairs on the thorax between the wing bases.

Similar species: All *Microdon* are superficially similar and require careful examination.

Observation tips: Larvae live in the nests of the Yellow Meadow Ant *Lasius flavus*. The majority of colonies are on chalk downland but there are also outlying colonies in North Wales and Norfolk. Adults are usually found basking on foliage, but occasionally visit flowers. Although not fully understood, this species appears to have very specific habitat requirements with respect to the location of anthills. It is also susceptible to changes in grassland management; for example, a switch from grazing to mowing has been known to wipe out a colony.

Microdon myrmicae / mutabilis[1] 

Wing length: 6·5–9 mm

Identification: The reddish-brown scutellum distinguishes these two species from other *Microdon*, but the adults cannot be separated. Although geographical location may provide a good clue, identification can only be confirmed by features of the mature larvae or puparium.

Similar species: All *Microdon* are superficially similar and require careful examination. Adult *M. myrmicae* and *M. mutabilis* cannot be separated.

Observation tips: The larvae of *M. myrmicae* live in the nests of the ant *Myrmica scabrinodis* in tussocks in wet pastures in western Britain and on the heathlands of southern England. It is scarce, but quite widespread. *M. mutabilis* larvae live in the nests of the ant *Formica lemani* and the only confirmed records are from northern Scotland – on Mull and near Inverness.

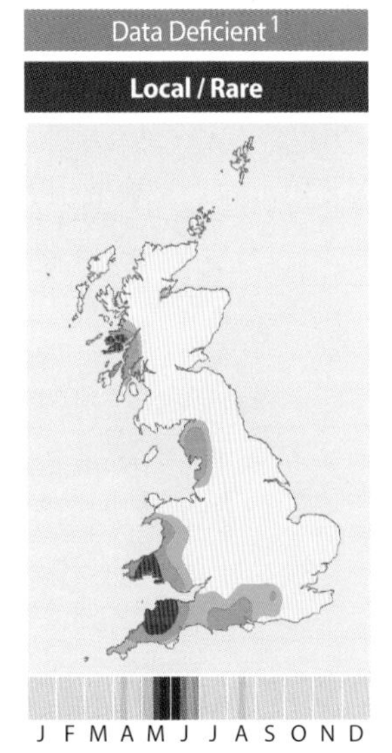

Microdon devius ♀ ×**6**

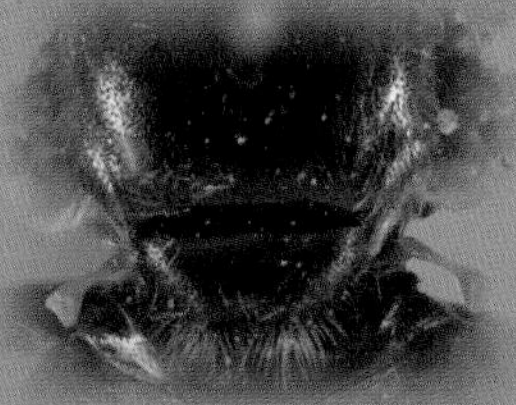

M. devius
scutellum **dark**

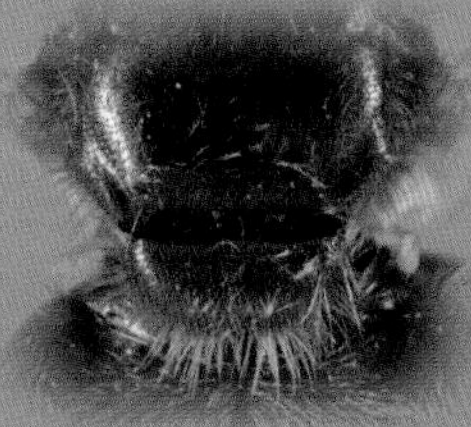

M. myrmicae/mutabilis
scutellum **reddish-brown**

Microdon myrmicae/mutabilis ♀ ×**6**

List of British and Irish hoverflies

The following table lists all 283 species known to occur in the British Isles at the time of writing (species are being added to the list at a rate of roughly one per year so it will probably go out-of-date quite quickly!). It also shows the classification into subfamilies and tribes. This is fairly controversial amongst taxonomists, but the taxonomy used in *British Hoverflies* (second edition) by Stubbs and Falk (2002) is followed here. The tribes in particular are useful in referring to features shared by groups of genera (see *pages 55–63*).

The 'Square count (**SC**)' status, gives the frequency of occurrence and the 'Difficulty of Identification (**ID**)' icons are as used in the species accounts (see *page 71*).

The 'Trend (**TR**)' column shows changes in the frequency with which the species has been recorded according to records submitted to the Hoverfly Recording Scheme from 1980 to 2010.

▲ indicates that the frequency of recording has increased significantly;
▼ indicates a significant decrease;
▬ indicates no significant change in recording frequency; and
- indicates that a trend could not be calculated.

Finally, the '**Status**' given in Ball & Morris (2014) is shown.

Page numbers are coded as follows:
Bold black text for species with full accounts
Bold grey text for species referenced in accounts with accompanying images
Light black text for species mentioned in accounts or in tables.

Chrysogaster solstitialis.

Name	SC	TR	Status	ID	Page
Sub-family: SYRPHINAE					

Tribe: BACCHINI

Name	SC	TR	Status	ID	Page
Baccha					
elongata (Fabricius, 1775)	Widespread	▼		👁	**74**
Melanostoma					
dubium (Zetterstedt, 1838)	Scarce	▼	Nationally Scarce	🔬	77
mellinum (Linnaeus, 1758)	Widespread	▬		🔍	**76**
scalare (Fabricius, 1794)	Widespread	▲		🔍	**76**
Platycheirus					
albimanus (Fabricius, 1781)	Widespread	▬		🔍	**80**
ambiguus (Fallén, 1817)	Frequent	▼		🔬	**80**
amplus Curran, 1927	Rare	–	Near Threatened	🔬	79
angustatus (Zetterstedt, 1843)	Widespread	▬		🔬	**86**
aurolateralis Stubbs, 2002	Rare	–		🔬	79
clypeatus (Meigen, 1822)	Widespread	▬		🔬	**86**
discimanus Loew, 1871	Local	▬	Nationally Scarce	🔬	79
europaeus Goeldlin, Maibach & Speight, 1990	Local	–		🔬	79
fulviventris (Macquart, 1827-8)	Frequent	▼		🔬	**88**
granditarsus (Forster, 1771)	Widespread	▬		👁	**88**
immarginatus (Zetterstedt, 1849)	Local	▼	Nationally Scarce	🔬	79
manicatus (Meigen, 1822)	Widespread	▼		🔍	**82**
melanopsis Loew, 1856	Scarce	▼	Near Threatened	🔬	79
nielseni Vockeroth, 1990	Frequent	–		🔬	79
occultus Goeldlin, Maibach & Speight, 1990	Frequent	–		🔬	79
peltatus (Meigen, 1822)	Widespread	▬		🔬	**84**
perpallidus Verrall, 1901	Local	▼	Nationally Scarce	🔬	79
podagratus (Zetterstedt, 1838)	Frequent	▬		🔬	79
ramsarensis Goeldlin, Maibach & Speight, 1990	Frequent	–		🔬	79
rosarum (Fabricius, 1787)	Widespread	▬		👁	**90**
scambus (Staeger, 1843)	Frequent	▼		🔬	79
scutatus (Meigen, 1822)	Widespread	▼		🔬	**84**
splendidus Rotheray, 1998	Local	–		🔬	79
sticticus (Meigen, 1822)	Local	–	Nationally Scarce	🔬	79
tarsalis (Schummel, 1836)	Frequent	▼		🔍	**82**
Xanthandrus					
comtus (Harris, 1780)	Frequent	▬		👁	**90**

Tribe: PARAGINI

Name	SC	TR	Status	ID	Page
Paragus					
albifrons (Fallén, 1817)	Rare	–	Critically Endangered	🔬	93
constrictus Simic, 1986	Ireland only	–		🔬	93
haemorrhous Meigen, 1822	Widespread	▲		🔬	**92**
tibialis (Fallén, 1817)	Scarce	▬	Near Threatened	🔬	93

Name	SC	TR	Status	ID	Page
Tribe: SYRPHINI					
Chrysotoxum					
arcuatum (Linnaeus, 1758)	Frequent	▼		🔍	**100**
bicinctum (Linnaeus, 1758)	Widespread	▲		👁	**98**
cautum (Harris, 1776)	Frequent	▼		🔍	**100**
elegans Loew, 1841	Local	▬	Nationally Scarce	🔬	**102**
festivum (Linnaeus, 1758)	Frequent	▬		👁	**98**
octomaculatum Curtis, 1837	Rare	–	BAP; Endangered	🔬	100
vernale Loew, 1841	Rare	–	Endangered	🔬	100
verralli Collin, 1940	Frequent	▬		🔬	**102**
Dasysyrphus					
albostriatus (Fallén, 1817)	Widespread	▼		👁	**116**
friuliensis van der Goot, 1960	Scarce	–		🔬	118
hilaris (Zetterstedt, 1843)	Scarce	–		🔬	118
pauxillus (Williston, 1887)	Rare	–		🔬	118
pinastri (De Geer, 1776)	Frequent	▼		🔬	**118**
tricinctus (Fallén, 1817)	Widespread	▼		👁	**116**
venustus (Meigen, 1822)	Widespread	▼		🔍	**118**
Didea					
alneti (Fallén, 1817)	Rare	–	vagrant	🔬	121
fasciata Macquart, 1834	Frequent	▼		🔬	**120**
intermedia Loew, 1854	Local	▬	Nationally Scarce	🔬	**120**
Doros					
profuges (Harris, 1780)	Scarce	▬	BAP; Near Threatened	👁	**106**
Epistrophe					
diaphana (Zetterstedt, 1843)	Frequent	▬		🔬	**142**
eligans (Harris, 1780)	Widespread	▲		👁	**144**
grossulariae (Meigen, 1822)	Widespread	▼		🔍	**142**
melanostoma (Zetterstedt, 1943)	Scarce	▬	Nationally Scarce	🔬	140
nitidicollis (Meigen, 1822)	Frequent	▬		🔬	**140**
ochrostoma (Zetterstedt, 1849)	Rare	–	Data Deficient	🔬	140
Episyrphus					
balteatus (Degeer, 1776)	Widespread	▲		👁	**138**
Eriozona					
syrphoides (Fallén, 1817)	Local	▼	Nationally Scarce	🔍	**112**
Eupeodes					
bucculatus (Rondani, 1857)	Local	▬		🔬	125
corollae (Fabricius, 1794)	Widespread	▬		👁	**126**
goeldlini Mazánek, Láska & Bicik, 1999	Rare	–		🔬	125
lapponicus (Zetterstedt, 1838)	Rare	–	vagrant	🔬	125
latifasciatus (Macquart, 1829)	Widespread	▬		🔍	**126**
lundbecki (Soot-Ryen, 1946)	Rare	–	vagrant	🔬	125
luniger (Meigen, 1822)	Widespread	▬		👁	**124**
nielseni Dušek & Láska 1976	Local	▼	Nationally Scarce	🔬	125
nitens (Zetterstedt, 1843)	Local	▬	Nationally Scarce	🔬	125
'species A'	Scarce	–		🔬	–

Name	SC	TR	Status	ID	Page
Leucozona					
glaucia (Linnaeus, 1758)	Widespread	▼		👁	114
laternaria (Muller, 1776)	Widespread	▬		👁	114
lucorum (Linnaeus, 17581	Widespread	▬		👁	112
Megasyrphus					
erraticus (Linnaeus, 1758)	Local	▼	Nationally Scarce	🔬	144
Melangyna					
arctica (Zetterstedt, 1838)	Frequent	▼		🔬	129
barbifrons (Fallén, 1817)	Scarce	▬	Near Threatened	🔬	129
cincta (Fallén, 1817)	Frequent	▼		🔍	132
compositarum (Verrall, 1873)	Frequent	▬		🔍	128
ericarum (Collin, 1946)	Rare	–	Vulnerable	🔬	129
labiatarum (Verrall, 1901)	Frequent	▬		🔍	128
lasiophthalma (Zetterstedt, 1843)	Widespread	▬		🔬	130
quadrimaculata (Verrall, 1873)	Local	▼		🔬	132
umbellatarum (Fabricius, 1794)	Frequent	▬		🔬	130
Meligramma					
euchromum (Kowarz, 1885)	Local	▼	Nationally Scarce	🔬	135
guttatum (Fallén, 1817)	Local	▼	Nationally Scarce	🔍	134
trianguliferum (Zetterstedt, 1843)	Frequent	▼		🔍	134
Meliscaeva					
auricollis (Meigen, 1822)	Widespread	▬		🔍	136
cinctella (Zetterstedt, 1843)	Widespread	▼		👁	136
Parasyrphus					
annulatus (Zetterstedt, 1838)	Local	▼		🔬	146
lineola (Zetterstedt, 1843)	Frequent	▼		🔬	146
malinellus (Collin, 1952)	Local	▼		🔬	146
nigritarsis (Zetterstedt, 1843)	Local	▬	Nationally Scarce	🔬	148
punctulatus (Verrall, 1873)	Widespread	▬		🔍	146
vittiger (Zetterstedt, 1843)	Frequent	▼		🔍	148
Scaeva					
albomaculata (Macquart, 1842)	Rare	–	vagrant	🔬	123
dignota (Rondani, 1875)	Rare	–	vagrant	🔬	123
mecogramma (Bigot, 1860)	Rare	–	vagrant	🔬	123
pyrastri (Linnaeus, 1758)	Widespread	▬		👁	122
selenitica (Meigen, 1822)	Frequent	▬		🔍	122
Sphaerophoria					
bankowskae Goeldlin, 1974	Rare	–	Data Deficient	🔬	108
batava Goeldlin, 1974	Local	▬		🔬	108
fatarum Goeldlin, 1989	Frequent	▬		🔬	108
interrupta (Fabricius, 1805)	Widespread	▬		🔬	108
loewi Zetterstedt, 1843	Scarce	–	Near Threatened	🔬	108
philanthus (Meigen, 1822)	Frequent	▼		🔬	108
potentillae Claussen, 1984	Rare	–	Vulnerable	🔬	108
rueppellii (Wiedemann, 1830)	Frequent	▼		🔬	110
scripta (Linnaeus, 1758)	Widespread	▲		👁	110
taeniata (Meigen, 1822)	Frequent	▬		🔬	108
virgata Goeldlin, 1974	Local	▬	Nationally Scarce	🔬	108
'species B' *sensu* Stubbs 1991	Scarce	–		🔬	–

Name	SC	TR	Status	ID	Page
Syrphus					
nitidifrons Becker, 1921	Rare	–		🔬	150
rectus (Osten Sacken, 1875)	Rare	–		🔬	150
ribesii (Linnaeus, 1758)	Widespread	▬		🔬	**150**
torvus Osten Sacken, 1875	Widespread	▬		🔬	**152**
vitripennis Meigen, 1822	Widespread	▬		🔬	**152**
Xanthogramma					
citrofasciatum (Degeer, 1776)	Scarce	▬		👁	**104**
pedissequum (Harris, 1776)	Widespread	▲		👁	**104**
stackelbergi Violovitsh, 1975	Local	–		🔬	104

Sub-family: ERISTALINAE

Tribe: CALLICERINI

Name	SC	TR	Status	ID	Page
Callicera					
aurata (Rossi, 1790)	Local	▬	Nationally Scarce	🔍	**156**
rufa Schummel, 1841	Scarce	▬	Nationally Scarce	👁	**154**
spinolae Rondani, 1844	Rare	–	BAP; Vulnerable	🔍	**156**

Tribe: CHEILOSIINI

Name	SC	TR	Status	ID	Page
Cheilosia					
ahenea von Roser, 1840	Rare	–	Vulnerable	🔬	–
albipila Meigen, 1838	Frequent	▼		🔍	**168**
albitarsis Meigen, 1822	Widespread	▬		👁	**174**
antiqua Meigen, 1822	Frequent	▼		🔬	170
barbata Loew, 1857	Scarce	▼	Nationally Scarce	🔬	166
bergenstammi Becker, 1894	Widespread	▬		🔬	**170**
caerulescens (Meigen, 1822)	Rare	–		🔍	**162**
carbonaria Egger, 1860	Local	▬	Nationally Scarce	🔬	170
chrysocoma (Meigen, 1822)	Local	▬	Nationally Scarce	👁	**174**
cynocephala Loew, 1840	Local	▬	Nationally Scarce	🔬	–
fraterna (Meigen, 1830)	Widespread	▼		🔬	**170**
griseiventris Loew, 1857	Frequent	–		🔬	164
grossa (Fallén, 1817)	Frequent	▬		🔍	**168**
illustrata (Harris, 1780)	Widespread	▬		👁	**162**
impressa Loew, 1840	Widespread	▲		🔍	**172**
lasiopa Kowarz, 1885	Frequent	▼		🔬	164
latifrons (Zetterstedt, 1838)	Frequent	▬		🔬	–
longula (Zetterstedt, 1838)	Frequent	▼		🔬	**166**
mutabilis (Fallén, 1817)	Local	▼	Nationally Scarce	🔬	174
nebulosa Verrall, 1871	Local	▬	Nationally Scarce	🔬	168
nigripes (Meigen, 1822)	Scarce	▬	Nationally Scarce	🔬	174
pagana (Meigen, 1822)	Widespread	▬		🔍	**166**
proxima (Zetterstedt, 1843)	Widespread	▬		🔍	**172**
psilophthalma Becker, 1894	Rare	–	Data Deficient	🔬	–
pubera (Zetterstedt, 1838)	Local	▬	Nationally Scarce	🔬	–
ranunculi Doczkal, 2000	Frequent	–		👁	**174**
sahlbergi Becker, 1894	Rare	–	Vulnerable	🔬	–
scutellata (Fallén, 1817)	Frequent	▬		🔍	**166**

Name	SC	TR	Status	ID	Page
semifasciata Becker, 1894	Scarce	▬	Near Threatened	🔬	**24**
soror (Zetterstedt, 1843)	Frequent	▲		🔬	166
urbana (Meigen, 1822)	Frequent	▼		🔬	–
uviformis (Becker, 1894)	Rare	–	Data Deficient	🔬	–
variabilis (Panzer, 1798)	Widespread	▬		🔍	**164**
velutina Loew, 1840	Local	▼	Nationally Scarce	🔬	172
vernalis (Fallén, 1817)	Widespread	▬		🔬	–
vicina (Zetterstedt, 1849)	Frequent	▼		🔬	170
vulpina (Meigen, 1822)	Frequent	▬		🔬	**164**
'species B'	Rare	–	Data Deficient	🔬	–
Ferdinandea					
cuprea (Scopoli, 1763)	Widespread	▲		👁	**176**
ruficornis (Fabricius, 1775)	Local	▬	Nationally Scarce	🔬	177
Portevinia					
maculata (Fallén, 1817)	Frequent	▬		👁	**176**
Rhingia					
campestris Meigen, 1822	Widespread	▬		👁	**178**
rostrata (Linnaeus, 1758)	Frequent	▲		🔍	**178**

Tribe: CHRYSOGASTRINI

Name	SC	TR	Status	ID	Page
Brachyopa					
bicolor (Fallén, 1817)	Scarce	▬	Nationally Scarce	🔬	195
insensilis Collin, 1939	Frequent	▬		🔬	**194**
pilosa Collin, 1939	Local	▼	Nationally Scarce	🔬	195
scutellaris RobineauDesvoidy, 1844	Frequent	▲		🔬	**194**
Chrysogaster					
cemiteriorum (Linnaeus, 1758)	Frequent	▼		🔬	**188**
solstitialis (Fallén, 1817)	Widespread	▬		🔍	**188**
virescens Loew, 1854	Frequent	▼		🔬	189
Hammerschmidtia					
ferruginea (Fallén, 1817)	Rare	–	BAP; Endangered	👁	**196**
Lejogaster					
metallina (Fabricius, 1777)	Widespread	▼		🔬	**192**
tarsata (Megerle in Meigen, 1822)	Local	▬		🔍	**192**
Melanogaster					
aerosa (Loew, 1843)	Frequent	▼		🔬	187
hirtella Loew, 1843	Widespread	▼		🔬	**186**
Myolepta					
dubia (Fabricius, 1805)	Local	▬	Nationally Scarce	🔍	**197**
potens (Harris, 1780)	Rare	–	BAP; Critically Endangered	🔬	197
Neoascia					
geniculata (Meigen, 1822)	Frequent	▬		🔬	183
interrupta (Meigen, 1822)	Local	▬	Nationally Scarce	🔬	183
meticulosa (Scopoli, 1763)	Frequent	▼		🔬	**182**
obliqua Coe, 1940	Frequent	▼		🔬	183
podagrica (Fabricius, 1775)	Widespread	▼		🔬	**182**
tenur (Harris, 1780)	Widespread	▼		🔬	**182**

Name	SC	TR	Status	ID	Page
Orthonevra					
brevicornis Loew, 1843	Frequent	▼		🔬	191
geniculata Meigen, 1830	Local	▼		🔬	191
intermedia Lundbeck, 1916	Rare	–		🔬	191
nobilis (Fallén, 1817)	Frequent	▼		🔍	**190**
Riponnensia					
splendens (Meigen, 1822)	Widespread	▼		🔍	**190**
Sphegina					
clunipes (Fallén, 1816)	Widespread	▬		🔬	**184**
elegans Schummel, 1843	Frequent	▼		🔬	**185**
sibirica Stackelberg, 1953	Frequent	▬		🔍	**184**
verecunda Collin, 1937	Frequent	▲		🔬	**185**

Tribe: ERISTALINI

Name	SC	TR	Status	ID	Page
Anasimyia					
contracta Claussen & Torp, 1980	Frequent	▬		🔍	**212**
interpuncta (Harris, 1776)	Scarce	▬	Nationally Scarce	🔬	213
lineata (Fabricius, 1787)	Frequent	▼		👁	**212**
lunulata (Meigen, 1822)	Local	▬	Nationally Scarce	🔬	213
transfuga (Linnaeus, 1758)	Frequent	▼		🔬	213
Eristalinus					
aeneus (Scopoli, 1763)	Frequent	▬		🔍	**208**
sepulchralis (Linnaeus, 1758)	Widespread	▼		🔍	**208**
Eristalis					
abusiva Collin, 1931	Frequent	▼		🔬	201
arbustorum (Linnaeus, 1758)	Widespread	▬		🔍	**202**
cryptarum (Fabricius, 1794)	Rare	–	BAP; Critically Endangered	🔍	201
horticola (Degeer, 1776)	Widespread	▼		🔍	**204**
intricaria (Linnaeus, 1758)	Widespread	▬		👁	**206**
nemorum (Linnaeus, 1758)	Widespread	▬		🔍	**202**
pertinax (Scopoli, 1763)	Widespread	▲		👁	**200**
rupium Fabricius, 1805	Frequent	▼		🔬	**204**
similis (Fallén, 1817)	Rare	–	vagrant?	🔬	201
tenax (Linnaeus, 1758)	Widespread	▲		👁	**206**
Helophilus					
affinis Wahlberg, 1844	Rare	–	vagrant	🔬	215
groenlandicus (Fabricius, 1780)	Rare	–	Data Deficient	🔬	215
hybridus Loew, 1846	Widespread	▬		🔍	**214**
pendulus (Linnaeus, 1758)	Widespread	▲		🔍	**214**
trivittatus (Fabricius, 1805)	Widespread	▲		🔍	**216**
Lejops					
vittatus (Meigen, 1822)	Scarce	–	Near Threatened	🔍	**216**
Mallota					
cimbiciformis (Fallén, 1817)	Local	▬	Nationally Scarce	🔍	**210**
Myathropa					
florea (Linnaeus, 1758)	Widespread	▲		👁	**210**
Parhelophilus					
consimilis (Malm, 1863)	Local	▼	Nationally Scarce	🔬	219
frutetorum (Fabricius, 1775)	Frequent	▼		🔍	**218**
versicolor (Fabricius, 1794)	Frequent	▼		🔍	**218**

Name	SC	TR	Status	ID	Page
Tribe: MERODONTINI					
Eumerus					
funeralis Meigen, 1822	Frequent	▼			220
ornatus Meigen, 1822	Frequent	▼			220
sabulonum (Fallén, 1817)	Scarce	—	Nationally Scarce		222
sogdianus (Stackelberg, 1952)	Rare	–	vagrant		221
strigatus (Fallén, 1817)	Frequent	▼			220
Merodon					
equestris (Fabricius, 1794)	Widespread	—			222
Psilota					
anthracina Meigen, 1822	Scarce	—	Nationally Scarce		224
Tribe: PELECOCERINI					
Pelecocera					
caledonicus Collin, 1940	Rare	–	Vulnerable		227
scaevoides (Fallén, 1817)	Scarce	▼	Nationally Scarce		226
tricincta Meigen, 1822	Scarce	▲	Nationally Scarce		226
Tribe: PIPIZINI					
Heringia					
brevidens (Egger, 1865)	Scarce	—	Nationally Scarce		235
heringi (Zetterstedt, 1843)	Frequent	—			234
latitarsis (Egger, 1865)	Scarce	—	Nationally Scarce		235
pubescens Delucchi & PschornWalcher, 1955	Local	—	Nationally Scarce		234
senilis Sack, 1938	Rare	–			235
verrucula (Collin, 1931)	Scarce	—	Data Deficient		235
vitripennis (Meigen, 1822),	Frequent	—			234
Pipiza					
austriaca Meigen, 1822	Frequent	—			230
fasciata Meigen, 1822	Local	▼			231
festiva Meigen, 1822	Ireland only	–			231
lugubris (Fabricius, 1775)	Local	▼	Nationally Scarce		231
luteitarsis Zetterstedt, 1843	Frequent	▼			232
noctiluca (Linnaeus, 1758)	Widespread	▼			232
notata Meigen, 1822	Frequent	▼			231
Pipizella					
maculipennis (Meigen, 1822)	Scarce	—	Nationally Scarce		237
viduata (Meigen, 1822)	Widespread	—			236
virens (Fabricius, 1805)	Frequent	—			236
Trichopsomyia					
flavitarsis (Meigen, 1822)	Frequent	▼			238
lucida (Meigen, 1822)	Rare	–			239
Triglyphus					
primus Loew, 1840	Local	—	Nationally Scarce		238
Tribe: SERICOMYIINI					
Sericomyia					
lappona (Linnaeus, 1758)	Frequent	—			242
silentis (Harris, 1776)	Widespread	—			242
superbiens (Müller, 1776)	Frequent	▼			240

Name	SC	TR	Status	ID	Page
Tribe: VOLUCELLINI					
Volucella					
bombylans (Linnaeus, 1758)	Widespread	▲		👁	**244**
inanis (Linnaeus, 1758)	Frequent	▲		👁	**248**
inflata (Fabricius, 1794)	Frequent	▬		👁	**246**
pellucens (Linnaeus, 1758)	Widespread	▬		👁	**246**
zonaria (Poda, 1761)	Frequent	▲		👁	**248**
Tribe: XYLOTINI					
Blera					
fallax (Linnaeus, 1758)	Rare	–	Critically Endangered	👁	**252**
Brachypalpoides					
lentus (Meigen, 1822)	Frequent	▼		👁	**254**
Brachypalpus					
laphriformis (Fallén, 1816)	Local	▬		🔍	**256**
Caliprobola					
speciosa (Rossi, 1790)	Rare	–	Near Threatened	👁	**254**
Chalcosyrphus					
eunotus (Loew, 1873)	Scarce	▲	Nationally Scarce	🔍	**258**
nemorum (Fabricius, 1805)	Frequent	▬		🔍	**258**
Criorhina					
asilica (Fallén, 1816)	Frequent	▬		🔍	**260**
berberina (Fabricius, 1805)	Widespread	▼		🔍	**262**
floccosa (Meigen, 1822)	Frequent	▬		🔍	**262**
ranunculi (Panzer, 1804)	Frequent	▲		👁	**260**
Pocota					
personata (Harris, 1780)	Scarce	▬	Nationally Scarce	👁	**264**
Syritta					
pipiens (Linnaeus, 1758)	Widespread	▬		👁	**266**
Tropidia					
scita (Harris, 1780)	Frequent	▬		👁	**266**
Xylota					
abiens Meigen, 1822	Local	▼	Nationally Scarce	🔬	268
florum (Fabricius, 1805)	Frequent	▼		🔬	268
jakutorum Bagachanova, 1980	Frequent	▲		🔍	**268**
segnis (Linnaeus, 1758)	Widespread	▬		👁	**270**
sylvarum (Linnaeus, 1758)	Widespread	▼		👁	**270**
tarda Meigen, 1822	Local	▼	Nationally Scarce	🔬	268
xanthocnema Collin, 1939	Local	▼	Nationally Scarce	🔬	**268**

Sub-family: MICRODONTINAE					
Microdon					
analis (Macquart, 1842)	Local	▬	Nationally Scarce	🔍	**272**
devius (Linnaeus, 1761)	Scarce	▬	Near Threatened	🔍	**274**
mutabilis (Linnaeus, 1758)	Rare	–	Data Deficient	🔍	**274**
myrmicae Schonrogge et al., 2002	Local	▲		🔍	**274**

Photographing hoverflies

Cameras

Digital cameras can be divided into three categories: **Compact cameras, Digital Single Lens Reflex (DSLR) cameras** and **Bridge cameras.**

Compact cameras are the small, point-and-shoot, pocket cameras. They often have a 'macro mode' to tackle remarkably small targets and are capable of producing very good images, but do have disadvantages that make them less suitable for hoverflies. Instead of a viewfinder, most current models have an LCD panel on the back that shows what the lens is 'seeing'. In bright, outdoor conditions it is often difficult to see this image clearly enough to be able to compose the photograph and ensure it is in focus. These cameras tend to be highly automated and give the user limited scope to control how the picture is taken. There may be an option to control the area on which the camera's autofocus system works. This is often presented as a rectangle, displayed on the LCD screen, which you can manoeuvre over the area on which you want to focus. If this facility is available it is usually best to focus on the insect's eyes. Finally, you will generally have to get the camera very close (*i.e.* a few centimetres) to a hoverfly to get an image that reasonably fills the frame. This can be quite a challenge and demands pretty good fieldcraft! Such cameras are increasingly being built into mobile phones, but are largely unsuitable for insect photography because they generally lack close-up modes. However, the technology is moving fast, so close-up modes may well be offered in future.

Focusing on the head of the hoverfly *Myathropa florea* using the LCD screen of a compact camera.

DSLR cameras are ideal for insect photography, but are expensive, bulky and require considerable knowledge and experience to use effectively. The big advantages are that the viewfinder allows you to see exactly what the lens is seeing, and that the lenses are interchangeable, allowing the use of gear specifically designed for small subjects. DSLRs also give you full control, so you can get the exposure and composition you want – providing you know what you are doing!

Bridge cameras are intermediate between a compact camera and a DSLR. They usually have an electronic viewfinder through which you see a small LCD screen, rather than actually looking through the lens. This works well in bright daylight. These cameras usually have a fixed lens, but this is often a 'super-zoom' that offers a very wide range of focal lengths with good macro capabilities. Physically, they are also intermediate: not as small as a compact, but not as heavy or bulky as a DSLR.

Magnification

The Drone Fly, *Eristalis tenax*, pictured overleaf, was about 15 mm long in life. In the picture, as printed here, it appears about 38 mm long (*i.e.* magnified about 2½×). But it could have been printed so that it filled the whole page – in which case it would appear about 110 mm long (*i.e.* magnified 7×). It is easy to see that 'magnification' depends upon the way in which the image is viewed and is, in consequence, not a very useful measure. A much more useful way to think about it is the size of the image on the camera's sensor.

A 'macro photo' is one in which the image of the subject is at least life-size on the camera's sensor – as shown. The term 'reproduction ratio' is used to describe the scale of the image – a ratio of 1:1 meaning life-size – and is a much more useful way to think about magnification. Successful hoverfly photographs are generally taken at a reproduction ratio between about half life-size (1:2) and 2 or 3 times life-size (2:1 or 3:1).

The Drone Fly, *Eristalis tenax*. A life-size image (reproduction ratio 1:1) and the way this would appear on an APS-C sensor used in some DSLRs.

There are several ways of achieving magnification:

Supplementary close-up lenses are screwed into the filter mount at the front of the lens. The magnifying power of such lenses is measured in dioptres. If your camera has a fixed lens then this is really the only possibility. Simpler compact cameras lack a screw thread around the front of the lens mount, but a push fit, or even some adhesive tape, can always be contrived. The disadvantage of close-up lenses is that the extra layers of glass in front of the lens mean some degradation in the quality of the image is likely – especially if using cheaper options.

When you focus a camera on a close object, the front of the lens will move farther out as you turn the focusing ring. How close you can focus is determined by how far it can move. If the camera has interchangeable lenses then the lens can be moved farther away from the camera by inserting **extension tubes**. Modern cameras have autofocus and exposure systems that require electrical connections between the camera and lens. Extension tubes must maintain these connections. Optics dictate that, for a lens focused at infinity, an extension equal to the focal length is needed to produce an image with a reproduction ratio of 1:1. Since the 'standard lens' supplied with many DSLRs usually has a focal length of around 50 mm, extension tubes tend to come in sets of about that length, typically supplied as three tubes of different sizes so that a number of combinations are possible. Extension tubes are just an empty tube, and therefore have no effect on image quality. However, moving the lens farther from the sensor spreads the light coming through it over a larger area, so the image is dimmer; the use of extension tubes to get to a 1:1 image dims the image by 2 stops (see below for an explanation of what this means).

Macro lenses are specifically designed to perform best when the subject is extremely close. They are very convenient and produce superb results, but they are expensive. It may be advisable to buy a moderately priced set of extension tubes first and experiment with those before advancing to a macro lens.

Extension tubes

Exposure

Cameras have three controls that affect the exposure:

The **shutter speed** is the length of time for which light is allowed into the camera, measured as a fraction of a second (*e.g.* 1/16, 1/250, *etc.*). The longer the shutter is open, the more light falls on the camera's sensor, making for a stronger signal. Since hoverflies are active and fast-moving animals, a short exposure to freeze motion is an advantage. Also, the camera must be kept still during the exposure to avoid 'camera shake' – where the whole picture is blurred because the camera moved during the exposure. So an exposure that is as fast as possible is desirable to stop the movement of both the subject and the camera.

Camera shake. This whole shot of *Baccha elongata* is blurred and the highlights extend as streaks.

The **aperture** is the size of the hole through which light enters the camera. It is measured in f-stops (*e.g.* F6·3, F16, *etc.*). (NB. the smaller the hole, the bigger the f-stop number – so F16 denotes a smaller aperture than F8.) A larger aperture means a brighter image on the sensor and hence a stronger signal. However, the size of the aperture also affects the distance over which focus can be attained (termed the **depth of field** – DoF) – which increases as the size of the aperture decreases (and as the f-stop number increases). This is important when photographing insects for identification purposes, where the picture needs to be in focus.

Insufficient depth of field. The thorax of this *Hammerschmidtia* is in focus, but the head is not because the DoF is too small. Pictures of animals are rarely acceptable if the eyes are not sharp!

Photo of *Syrphus ribesii* drinking taken at high ISO (1,600). The magnified insert shows the noise visible in the dark areas.

Finally, the **sensitivity** of the camera's image sensor can be changed. This is measured by an ISO number. Lower ISO numbers (*e.g.* ISO 100) indicate less sensitivity – more light is needed to form a usable image. As the ISO number is raised (say to ISO 400 or 800), the amplification in the imaging system's electronics increases and an image can be recorded with less light. Some modern DSLRs will allow very high ISO settings (up to about ISO 25,600). So why not always use the highest ISO setting your camera provides? The problem is image 'noise'. This appears as a random scatter of lighter or darker pixels in the image. It is particularly obvious as speckling in dark areas. If you turn up the ISO too far, the picture will be degraded by noise. So try to keep to low ISO settings for noise-free images.

When you photograph a hoverfly, there is only so much light available. The ideal is a fast shutter speed to stop motion, a small aperture to give as much DoF as possible and a low ISO setting to avoid noise. In reality there is rarely enough light to achieve all of these at the same time (otherwise it would be too easy!), so compromise is needed.

Each of the scales of measurements is calibrated in halving or doubling of the amount of light. So a shutter speed of 1/125 lets in about half as much light as 1/60. Similarly, changing the f-stop by one increment halves or doubles the area of the aperture – F8 lets in twice as much light as F11. Equally, you need twice as much light to form an image at ISO 100 as you would at ISO 200. For example, an exposure of 1/125th second at F8 using ISO 200 gives a correctly exposed image, the same exposure would be obtained with an exposure of 1/60 at F11 (shutter open for twice as long, but the aperture half the size) or the ISO could be doubled to 400 and use 1/125 at F11 (half the size of the aperture, but double the sensitivity of the sensor). One of the skills of a photographer is juggling these settings to get the desired picture.

A point-and-shoot camera (or a DSLR in 'Auto' mode) will make the decision for you. Modern cameras can make a pretty good job of taking 'typical' pictures. However, automatic metering systems may not cope well with close-up photography. Point-and-shoot cameras often offer a range of picture styles, 'close-up' (often denoted by a flower icon) usually being one of the options. This setting helps the metering system and increases the chance of a good shot. So use the close-up setting if your camera offers it.

More sophisticated cameras allow you to take control. Hoverfly pictures on a DSLR will most likely be taken using 'Aperture Priority' mode. In this mode, you set the aperture (to achieve the necessary DoF) and the camera's meter will suggest a shutter speed appropriate for the available light. You must decide whether that shutter speed is fast enough and, if not, adjust the ISO and/or aperture to increase the shutter speed.

Electronic flash

Electronic flash provides a way out of the dilemma in which a fast shutter speed is needed to stop motion but a small aperture is required to achieve sufficient DoF. Small flash guns provide a bright source of light that can be positioned close to the subject, thereby allowing a small aperture to be used and giving a greater DoF. The way in which an electronic flash gun works ensures that the flash of light it produces is very brief indeed – typically 1/1,000th of a second or less in duration. This is short enough to counter camera shake and stop most motion. (But, whilst hovering, a hoverfly typically beats its wings around 300 times per second. In 1/1,000th of a second it will therefore go through 1/3rd of a wing-beat – and as a result the wings will still show motion blur!)

Electronic flash can, however, result in a brightly lit subject against a dark, or even completely black, background because the flash does not reach far enough to illuminate the background. This is not a problem when photographing a hoverfly resting on a flower or leaf that fills the whole field of view. The harshness of the lighting can also be ameliorated by fitting the flashes with suitable diffusers.

Another use for electronic flash is to fill in shadows in bright sunshine. A hoverfly sitting in full sun will have a strong, dark shadow. A bit of flash to soften the shadows can make for a much better picture.

Two pictures of male *Helophilus pendulus* taken in the same location with a 100 mm macro lens at around 1:1 and at F11. The picture on the left was taken in bright daylight (1/160, 400 ISO) – note the strong shadows. The picture on the right was taken using the twin flash set-up shown below (1/250, 100 ISO) – note how the flashes have filled in for each other resulting in little shadowing whilst the colours are stronger and more contrasty. The background (a yellow, polythene bowl) completely fills the frame in the right-hand shot, so there is no problem with a black background.

Support

Another way of avoiding camera shake and allowing slower shutter speeds is to support the camera using a tripod, monopod or beanbag. Longer focal length macro lenses and close-up flash can get rather heavy, so extra support is all the more necessary.

Macro lens and close-up flash set-up

Tripods are awkward to use when stalking hoverflies, but can be erected where you expect flies to appear (*e.g.* next to a patch of flowers). You can trigger the camera using a remote release when the subject comes into the pre-set field of view. Another possibility is to set the autofocus system to trigger the shutter when a subject moves into focus (although additional software such as Magic Lantern www.magiclantern.fm may be required). This is one way of tackling shots of hoverflies in flight. A monopod is much easier to use than a tripod when stalking and provides support while still enabling you to focus by rocking backwards and forwards.

A beanbag placed on the ground can support the camera when you need to get very low down. In this situation, it can be rather difficult to get your head in position to see through the viewfinder – but can be overcome with a right-angled attachment fitted on the viewfinder. Some cameras have an articulated LCD screen on the back, which is an advantage when composing and focusing in these circumstances.

Fieldcraft

An insect the size of a hoverfly usually requires around a life-size (1:1) image to fill the camera's frame – a little less for large species like *Volucella*, but perhaps twice or even three times life-size for small species like *Melanostoma*, *Paragus* or *Pipizella*. Achieving such magnification requires close approach, although the actual distance depends upon the focal length of the lens you are using (the longer the focal length, the farther away it will focus to life-size).

Getting close to a live hoverfly in the field can be a challenge but the following tips should help you to get the photographs you desire:

- Approach slowly and smoothly, avoiding sudden movements – hoverflies are very sensitive to movement.
- Do not let your shadow fall over the fly – that is a sure way of scaring it off!
- Be very careful not to disturb the vegetation on which the fly has settled.
- If a hoverfly is working its way over a patch of flowers or a large umbel, then position yourself at the nearest convenient spot and wait for the fly to come to you. Males of many species will return repeatedly to a perch when they are searching for females. Position yourself close to one of these perches and be patient.
- Hoverflies rarely show much reaction to flash, but may respond to the infra-red beam sometimes used by compact and bridge cameras to assist autofocus.
- When you are working at around 1:1, turn the autofocus off and focus by gently rocking back and forth. Autofocus systems are often fooled by stems and the like nearer and farther from the subject. This can lead to the camera 'seeking', where the focus mechanism keeps shifting from one point to another, which can lose you the shot!
- Do not try to bring the insect into focus and then freeze. Instead, try to move smoothly through the point of sharpest focus, pressing the shutter at the correct moment without pausing. If your camera can take multiple shots, take a short burst of, say, three. This both increases the chance that one will be taken at the optimum DoF, but also means that the slight jerk as the shutter button is depressed only affects the first shot.
- In general, it is best to aim to get the animal's eyes in focus, as you can afford to have wing-tips, the ends of the legs, *etc.* going out of focus without spoiling the picture. Remember that about half the DoF is in front of the point of sharpest focus and half behind. So choose the point where you focus with this in mind. (It is often stated that 1/3rd of the DoF is in front of the focal point and 2/3rds behind, but this is only true if the subject distance is much greater than the focal length of the lens).

Finally, it is important to think about the vegetation you may be trampling in order to get into position for a photograph. If there is any risk of damaging the vegetation, especially on protected sites (*e.g.* chalk grassland, tall fen, *etc.*) then stop – the welfare of the wildlife and habitats must come first!

Further reading

Read the manual and make sure you know how your camera works and can put it into close-up mode or adjust the ISO settings, *etc.*

Insect Photography: Art and Techniques (Crowood Press, 2012) is by John Bebbington who ran a course on the subject for the Field Studies Council for many years.

Two internet sites that seem to be well moderated and features articles by knowledgeable practitioners are Cambridge in Colour (www.cambridgeincolour.com) and The Luminous Landscape (www.luminous-landscape.com).

Catching hoverflies

Using a hand lens to examine a hoverfly

The key piece of equipment required for catching hoverflies is an insect net. These are available from a variety of entomological suppliers and are based on four-fold landing net frames sold to anglers. The bag is made out of a light, rip-stop material. Nets made for catching butterflies and moths tend to be black, but these are not suitable for flies as you cannot see them in the net! Dipterists mostly use white nets. The bag needs to be long enough so that the end folds over the edge of the frame, trapping the fly in a closed pocket.

These frames have a standard sized screw fitting and a variety of poles can be used interchangeably. They are typically sold with an aluminium pole about 1 m long. These are fine for most purposes, but a longer handle can be useful, especially early in the year, when trying to catch hoverflies visiting blossom such as willows and Blackthorn. Some species, such as *Cheilosia grossa*, *Criorhina ranunculi* and *Platycheirus discimanus*, tend to stay high up and out of reach – a telescopic landing net handle from a fishing tackle shop can extend to 3 m or more and allow you to reach the tops of bushes.

Hoverflies have very good reactions and can move extremely fast, so catching them can be a challenge. A stroke of the net that comes from behind and underneath is usually the most effective technique. Small species such as *Melanostoma* and *Platycheirus* are hard to spot and are often best found by 'sweeping'. This involves sweeping the net gently back and forth through grass and other low vegetation. Watch out for brambles, briars other thorny shrubs and barbed wire to avoid inadvertently shredding your net!

Individual hoverflies can be picked out between the thumb and fingers for close examination without harming them. However, if you want to keep the hoverfly for later examination, you need to transfer it to a container. Small, transparent specimen tubes (glass or polystyrene) with polythene caps are useful for this. Take the top off the tube and place it over the hoverfly in the net so that it is trapped in the tube against the netting. Then shake the fly into the tube and slide the cap on. This becomes fairly routine with practice, but it is inevitably the best catch that manages to escape!

The pooter is a better option for extracting smaller hoverflies, especially after you have been sweeping. There are several designs, but the principle is the same. There is a closed, transparent container with two tubes leading into it through a rubber bung. You suck through one and the fly is sucked into the container through the other. The tube you suck through has some gauze over the inner end so that you don't get a mouthful of flies!

A good way to empty your net after sweeping is to first carefully let out the more aggressive and larger catch, like bumblebees, and then to put the net over your head with the end of the bag held so that it is pointing towards the brightest part of the sky.

Using a long-handled net.

Using a pooter to extract flies from a standard net.

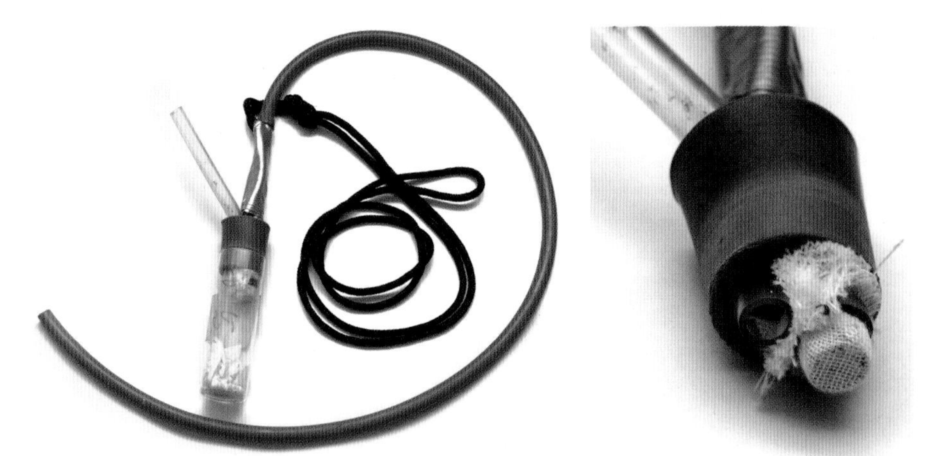

A pooter. Suck through the rubber tube and flies are sucked into the glass specimen bottle through the clear polythene inlet tube. The tube you suck through has gauze over it to avoid flies getting sucked into your mouth. The black cord is a lanyard to avoid losing or dropping the pooter.

Safety considerations:

- Don't poot insects collected off carrion, dung or fungi. You might breath in spores or unpleasant bacteria! (This is not usually an issue when catching hoverflies.).
- Make sure your pooter tube is cleaned and ventilated before re-use to avoid getting a mouthful of Ethyl Acetate fumes next time you use it.

Insects will tend to walk up the netting towards the light, so this keeps them away from the entrance and escape, and also away from your face. You then pick off the hoverflies by sucking them into the pooter.

Hoverflies can be killed by putting a couple of drops of Ethyl Acetate on a twist of tissue paper and dropping this into the container with them; the vapour knocks them out very quickly. It is advisable to use glass or polythene tubes for this purpose and to avoid putting Ethyl Acetate into polystyrene tubes as it will dissolve the plastic. Ethyl Acetate can be obtained from entomological suppliers, but cannot be sent by post. You will therefore either need to buy it in person at an entomological exhibition or have it sent by a courier service (which makes it rather expensive). An alternative is to put the tubes containing the hoverflies in the freezer for a couple of hours.

Curating specimens

Specimens of hoverflies are best kept dry and pinned using headless, stainless steel micro-pins. These are available from entomological suppliers and come in various thicknesses (designated by letters, where A is the thinnest) and lengths (designated by numbers: 1 = 10mm, 2 = 15mm, *etc.*). A1 pins are the shortest and finest and are suitable for pinning small hoverflies like *Platycheirus*; B2 pins are suitable for medium-sized species like *Syrphus*; and C3 pins are suitable for big hoverflies like *Volucella*. Small hoverflies are often pinned on their sides with the pin passing through the thorax. Ideally, the pin should be inserted slightly off the vertical so that it comes out in a slightly different place on the other side of the thorax. This means that any feature it penetrates (and potentially destroys) will survive on at least one side. Bigger hoverflies are often pinned flat with the pin passing from the top to the bottom of the thorax.

Hoverflies need to be pinned when fresh. If you allow your specimen to dry out it will become brittle and pinning it in this state risks parts falling off. Once the micro-pin has been inserted through the hoverfly, spread out its wings and legs so that everything can be seen and hold the various parts in place with other pins while the specimen dries out.

A male specimen of the small hoverfly *Sphaerophoria* that has been side-pinned using an A1 micro-pin and then staged and mounted on a continental pin with data and determination labels.

This is also the stage at which to tease out a male's genitalia so it is visible. Again, the genital capsule can be hinged out and held in place with a pin until it dries and will then stay in the position required. Once the specimen is dry, the temporary positioning pins can be removed.

For long-term storage, the micro-pinned specimen should be pinned to a 'stage'; this is a strip of vibration-absorbing material such as plastazote or polyporus (the Birch Bracket Fungus), which is mounted on a much longer pin. Continental length stainless steel pins from entomological suppliers are best for this purpose. The specimen should always be handled by the large pin to avoid damage and never directly touched or handled by the micro-pin.

The most important part of any specimen is the labels. These are also kept on the large pin and record the details about the specimen. There will normally be (at least) two labels: a data label, including where and when the specimen was caught; and a 'determination' label which gives the identification and the name of the person who identified it.

Why keep a collection?

There are a number of good reasons for keeping specimens of hoverflies in a collection.

Vouchers: You can prove to yourself and others that you got the identification correct. This is especially important when you are starting off. The best way to learn is to try naming some specimens and then to get them checked by an expert. Nevertheless, however expert you may become, it remains important to keep vouchers for difficult and uncertain identifications.

Comparison: The identification process often involves comparative judgements (*e.g.* bigger than …; eye hairs darker than …). In these cases, it is very useful to have material you have previously identified to hand so that you can remind yourself what the various options look like.

Species splits: Knowledge of hoverflies is growing all the time and there have been a number of recent taxonomic revisions which have led to species being 'split'. There are also further such revisions under consideration. When a species is split into two or more new species, it is only possible to confirm your previous identifications if you have kept specimens that can be re-examined.

One of the authors, Roger Morris, dealing with the day's catch during a Dipterists Forum summer field meeting.

A collection of voucher specimens in a plastazote-lined wooden store box.

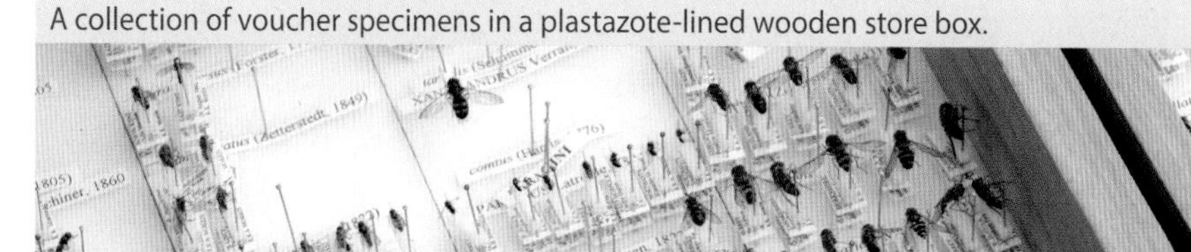

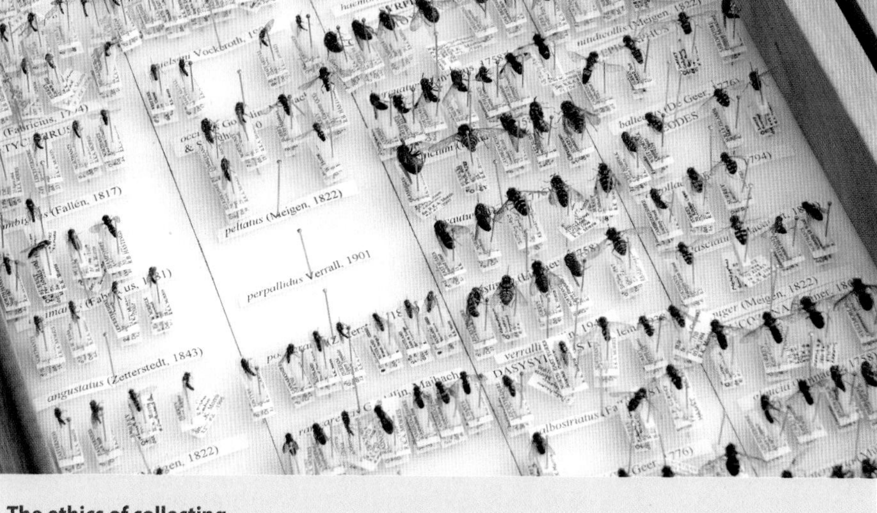

The ethics of collecting

Many naturalists frown upon the collection of specimens. The prevailing ethos is "leave nothing but footprints, take nothing but photographs". As we have seen above, this approach does not work for a large proportion of Diptera, or indeed for many other invertebrates. It is often necessary to examine a dead specimen rather closely in order to ensure an accurate identification.

So is there a potential conflict between collection and conservation? The authors of this book have both spent most of their working lives employed by conservation organisations and do not consider there to be a conflict. Whilst we recognise that a few rare species have such restricted breeding sites that collecting could be damaging, we believe that insect faunas are at far greater risk through not knowing which sites are important and how they should be managed. The collection of a few voucher specimens is extremely unlikely to be damaging to hoverfly populations and, in our view, responsible collecting that adds to our knowledge of the distribution and biology of hoverflies, should be encouraged. However, we would stress that if an animal has died to generate an identification then use should be made of the resulting record! It should be lodged with the Recording Scheme and subsequently made available to conservation organisations and researchers. Although not a good reason to condone collecting, it is worth bearing in mind that you will kill many more hoverflies on the front of your car driving to and from a site than you are ever likely to collect.

The Joint Committee for the Conservation of British Invertebrates (JCCBI) first published A Code of Conduct for Collecting Insects and Other Invertebrates in 1972 and this was most recently revised in 2002. It is available from http://www.royensoc.co.uk/InvLink/Index.html and anyone wishing to collect hoverflies should familiarise themselves with this Code and adhere to its principles.

Legislation and conservation

Only a few insect species are afforded some degree of legal protection in Britain, the principal legal instrument being the Wildlife and Countryside Act (1981). As a signatory to the Rio Convention on Biological Diversity in 1992, the UK government published its first Biodiversity Action Plan (BAP) in 1994. Species were selected for conservation action on the basis of conservation decline; highly localised distribution; international threat and importance and Red Data Book status. The UK Biodiversity Action Plan currently lists seven hoverfly species as priorities for conservation action: *Blera fallax*, *Callicera spinolae*, *Chrysotoxum octomaculatum*, *Doros profuges*, *Eristalis cryptarum*, *Hammerschmidtia ferruginea* and *Myolepta potens*.

Single-species studies have now been conducted on all seven UKBAP priority species, with varying degrees of success. Most success has been achieved with the Pine Hoverfly *Blera fallax* and the Aspen Hoverfly *Hammerschmidtia ferruginea*, whose ecology are now well known as a consequence of work by the Malloch Society and, latterly, through a PhD study undertaken by Ellen Rotheray. Both *B. fallax* and *H. ferruginea* have been shown to have extremely small populations and to be at considerable risk of extinction. However, both have been shown to respond well to active conservation management, with new breeding sites created by judicious use of the chainsaw to create new rot holes and fallen timber respectively. *H. ferruginea* is a 'flagship' species for a wider assemblage of insects associated with Aspen woods in Scotland. These woods are mostly small and many are heavily grazed, showing few signs of recruitment of new trees. They are potentially vulnerable to other conservation initiatives such as the re-introduction of European Beaver.

Useful population studies of *Eristalis cryptarum* have shown this species to be confined to a very small part of southern Dartmoor but the larvae have still not been found despite careful work over several years. The larvae of both *Myolepta potens* and *Callicera spinolae* are known to occur in water-filled rot holes but these species are still highly elusive and relatively little is known about their distribution or the factors that make them so restricted. *Doros profuges* and *Chrysotoxum octomaculatum* have proved to be particularly elusive and little new information has emerged.

Eristalis cryptarum.

Recording hoverflies

The Hoverfly Recording Scheme is one of the most active wildlife recording schemes in Britain, and welcomes data in computer-readable or paper formats from anyone. Recorders who use packages such as MapMate and Recorder can be readily accommodated. Otherwise, data are preferred in spreadsheet format that include the following basic headings: (red = required; green = optional).

Species name	Location name	OS Grid reference	Date	Recorder's name	Sex/ life stage	Number	Notes & comments

In addition, posts on media such as iSpot, forums on the Recording Scheme website and UK Hoverflies Facebook page are regularly followed up (for website addresses see *pages 303–304*).

Data assembled by the Recording Scheme are used in a wide variety of ways. Traditional distribution maps, which show the 10 km squares of the Ordnance Survey National Grid in which a species has been reported, are posted on the website (see *page 303*) and published in atlases. But data can be used in many more ways, including investigating trends in abundance, changes in distribution, changes in the flight time of species (phenology) and responses to changing climate. In addition, data from the Recording Scheme are being used in major studies of pollinators by various academic teams. Analysis of trends and phenology is highly dependent upon complete runs of data. All records are sought and not just records of 'interesting' species or the first/last date for a species in a year. Contributors are encouraged to maintain logs of everything they see and, if possible, to make counts. Records of even the commonest species can be important because these are potential bellwethers for changes in the countryside that might not otherwise be detected.

Long-term trend analysis also depends upon the robustness of the dataset; consequently, the Recording Scheme organisers scrutinise data submitted to them and may seek confirmation of the identity of trickier species. Records of species that are considered to be difficult to identify cannot be accepted without supporting evidence (*e.g.* a

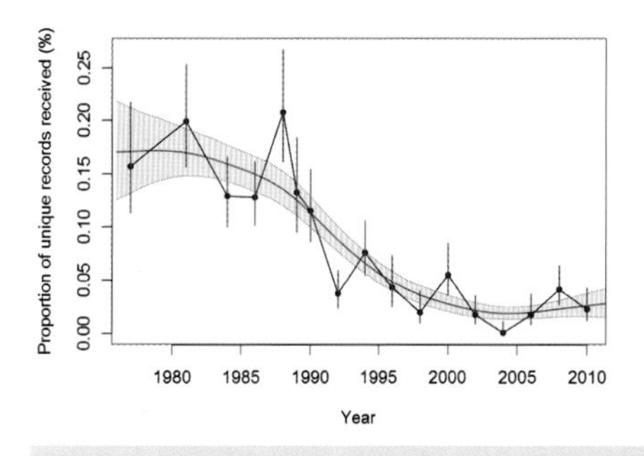

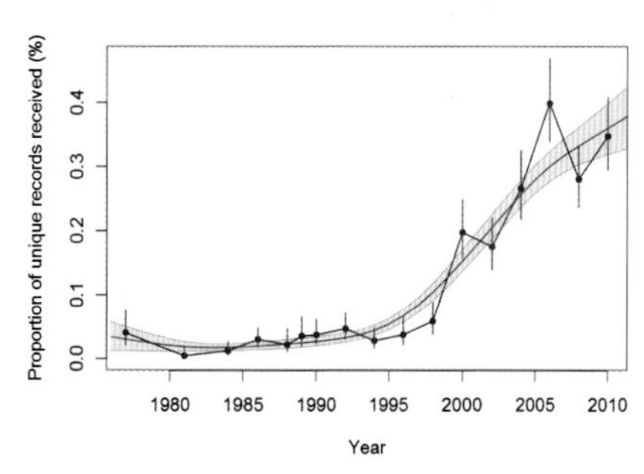

Trends in the frequency with which *Parasyrphus vittiger* (LEFT) and *Rhingia rostrata* (RIGHT) have been reported to the Hoverfly Recording Scheme. The plots show the proportion (%) of unique records received that were of the given species, together with a smoothed trend and confidence interval. See the Hoverfly Atlas (Ball *et al.* 2011) for a fuller explanation of the method used.

photograph and/or a dead specimen). An identification service is offered for specimens that require microscopic examination. On occasion, large batches of specimens can be identified if required for research projects.

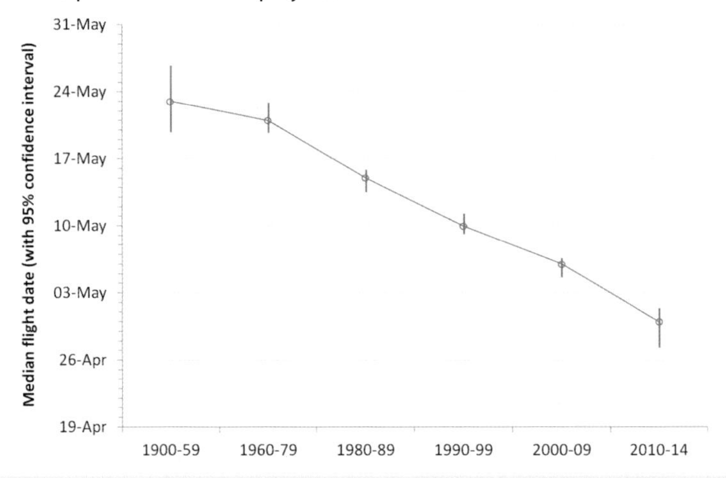

Changes in the flight period of *Epistrophe eligans*: data submitted to the recording scheme show that this hoverfly is on the wing earlier in the year than it used to be. The plot shows the median date of records for each period and its 95% confidence interval. This date is now about three weeks earlier than it was prior to 1980.

DEVELOPING IDENTIFICATION SKILLS

Identification skills take time to develop. New recorders typically start with the more frequent and conspicuous species and encounter a wider range of species as their interest develops. This growing interest also stimulates the development of new field skills as many hoverflies are not likely to be found without specifically searching for them. Although some species can be recognised by looking at pictures, it is often necessary to use keys, such as those in the main monographs (Stubbs & Falk, 2002 and van Veen, 2004 (see **Further reading**, *page 302*) to reach an identification. This requires detailed examination of a specimen under a hand lens or binocular microscope.

Identification requires an understanding of a whole new terminology describing the insect's body and its features. This book endeavours to explain and illustrate such terms, but there are many tips and tricks involving the handling, positioning and lighting of specimens that it is much easier to demonstrate than to describe in words. One way of developing these skills is to find somebody who is already experienced and get their help. Events run by local naturalists or recording groups (often organised in association with the Wildlife Trusts, local Natural History Societies, Local Record Centres or museums) are a good place to start when looking for such help. Training courses are especially beneficial and, even if they cover other insect groups, will introduce useful terminology and techniques. The Hoverfly Recording Scheme organisers provide a training service and regularly run weekend courses throughout Britain. Whilst the Recording Scheme is able to provide tutors, training material and handouts *etc.*, a local organiser is required to arrange a venue. The maximum group size is 12. Further details can be obtained via the Recording Scheme website (see *page 303*).

Gardening for hoverflies

Gardens can be great places to observe hoverflies – in one Peterborough garden alone nearly 70 species have been recorded. This may in fact not be an unusual number, but since relatively few people monitor their garden intensively it is not possible to be sure.

The usual way of thinking about the use of the garden by invertebrates is to focus on nectar sources. This is as true for hoverflies as it is for butterflies, although the choice of plants differs. Hoverflies will visit butterfly lures such as *Buddleja* and *Hebe* but they also benefit from a broader suite of flowers throughout the season.

There are some simple things that can be done to make a garden more attractive to hoverflies. For example, since hoverflies love *Allium* flowers it is worth leaving leeks that have bolted to flower. Alternatively, consider planting some of the bigger ornamental *Allium* species to see what they attract. Parsnip flowers also provide an attractive nectar source, so try allowing a few less vigorous plants to bolt. Fennel, as well as most native umbellifers, such as Hogweed and Angelica, are also potentially good lures. If you do not like the idea of encouraging them in your garden, try growing them in pots!

Ironically, one of the most attractive plants to hoverflies is one of the gardener's worst enemies – Ground Elder. Although it is invasive, if you have a big enough garden try not to eliminate entirely this fantastic nectar source. This illustrates an important point: many of the best hoverfly lures are wild flowers. As native shrubs are particularly attractive to hoverflies, consider planting a hedge of Blackthorn and/or Hawthorn, or growing dogwoods or Wild Privet (rather than its Japanese counterpart) as ornamental plants. Hoverflies will not confine their attention to native species and many ornamental plants can also provide good nectar sources. The list on *page 301* provides a few ideas and indicates the months in which they are usually in flower.

Many of the hoverflies that visit gardens have larvae that feed on aphids. It is therefore worth considering 'companion planting', an approach whereby flowers and vegetables are grown close together. The theory is that hoverflies attracted to the flowers' nectar

Marmalade Fly *Episyrphus balteatus* on Busy Lizzie.

sources lay their eggs on the nearby vegetables, where there are often aphid colonies. The resulting larvae feed on the aphids and provide a natural means of pest control.

Mature larva of *Epistrophe*.

Thriving populations of predacious species can also be encouraged by the provision of over-wintering sites for puparia. One good example is *Epistrophe eligans*, which is fond of aphids on plum trees. The larvae of this hoverfly need a dry, sheltered place on the ground to pupate, and benefit from some corrugated cardboard stuffed into a suitable waterproof receptacle that is open at one end (pointing downwards to avoid filling with rainwater).

A "nest box" for *Myathropa florea*. An empty plastic drink bottle with a hole cut in the side has been taped to a tree. The water in the bottom contains a handful of twigs and leaves.

Other garden features that are important for hoverflies could be maintained by avoiding modern silvicultural practices. For example, hoverflies that breed in decaying roots do not benefit from the action of the modern stump-grinder. Unfortunately, this tool is now widely used in many parks and gardens, resulting in the elimination of much prime hoverfly habitat.

Another way of encouraging hoverflies to inhabit your garden is to provide hoverfly nest boxes. An old bucket placed in a corner and filled with twigs and leaves, and topped up with water, will rapidly attract *Myathropa florea* whose larvae often occur in good numbers in such situations. For a more bespoke version, cut a hole in the side of a two-litre plastic drink bottle and tape it low

Garden hoverfly monitoring

Even if you do not go far from your garden, you can make an important contribution to monitoring hoverflies as there are currently only two long-term datasets available. The studies that contributed to these data sets were undertaken by Alan Stubbs in his Peterborough garden (Stubbs, 1991) and by Dr Jennifer Owen in her garden in Leicestershire (Owen, 2010). The technique used by Alan Stubbs is based on the concept of the butterfly transect methodology devised and used by Butterfly Conservation. It involves recording the hoverflies seen along a standardised walk around the garden each week from April to October. Jennifer Owen ran a malaise trap in her garden from 1971 to 2000. Maintaining a daily log of species and numbers seen is more helpful than maintaining a record for individual weeks.

down on the sheltered side of a tree (away from direct sunlight as this will cause water temperature to rise in the Summer and risks cooking the larvae). The bottle should be partly filled with sawdust, twigs and water and kept topped up during dry weather. Such artificial nests may also occasionally yield scarcer species than *M. florea*.

Garden plants that are attractive to hoverflies

January to March
Winter Aconites *Eranthis* spp.
Crocuses *Crocus* spp.
Mahonia spp.
(different species flower at different times from November through to March)
Viburnum tinus
Willows (sallows) *Salix* spp.

April/May
Cherry Laurel *Prunus laurocerasus*
Cherry Plum *Prunus cerasifera*
Mexican Orange Blossom *Choisya* spp.
Clematis spp.
(early flowering species)
Dandelions *Taraxacum* agg.
(leave them in the garden)
Spurges *Euphorbia* spp.
Forget-me-nots *Myosotis* spp.
Fruit tree blossom
Green Alkanet *Pentaglottis sempervirens*
Lithodora spp.
Pieris spp.
Rosemary *Rosmarinus officinalis*
Skimmia japonica
(male plants are best)
Tree Peony *Paeonia suffruticosa*
(single varieties)

May/June
Blackberry/Loganberry and **Raspberry** *Rubus* spp.
Pot Marigolds *Calendula* spp.
(single varieties)
Camellia spp. (single varieties)
Crane's-bills *Geranium* spp.
(herbaceous hardy Geraniums)
Dogwoods *Cornus* spp.
Feverfew *Tanacetum parthenium*
Marjoram *Origanum majorana*
Pieris spp.
Wild Privet *Ligustrum vulgare*
(not Japanese Privet!)

July/August
Yarrows *Achillea* spp.
Anise Hyssop *Agastache foeniculum*
Masterworts *Astrantia* spp.
Bishop's Flower *Ammi majus*
Shrubby Hare's-ear *Bupleurum fruticosum*
Borage *Borago officinalis*
Bronze Fennel *Foeniculum vulgare*
Buddleja davidii
Buddleja × weyeriana
Canadian Goldenrod *Solidago canadensis*
Oxeye Daisy *Leucanthemum vulgare*
Sea-hollies *Eryngium* spp.
Hebe spp.
Japanese Anemone *Anemone hupehensis*
Knapweeds *Centaurea spp.*
Lavenders *Lavandula spp.*
Leeks *Allium* spp.
(leave a few to run to flower)
Chinese Privet *Ligusticum lucidum*
Marjoram *Origanum majorana*
Mexican Sunflower *Tithonia diversifolia*
Michaelmas Daisy *Aster amellus*
Mints (and relatives) *Mentha* spp.
Poppies *Papaver* spp.
Scabiouses *Scabiosa* spp.
Shasta Daisy *Leucanthemum × superbum*
Verbena bonariensis
Western Pearly Everlasting *Anaphalis margaritacea*

September/October
White [or Heath] **Aster** *Symphyotrichum* [=*Aster*] *ericoides*
Symphyotrichum [=*Aster*] *lateriflorum* [=*lateriflorus*]
Tree-ivy × *Fatshedera lizei*
Japanese Aralia (or Japanese Fatsia) *Fatsia japonica*
Helenium spp.
Hemp Agrimony *Eupatorium cannabinum*
Common Ivy *Hedera helix*
Coneflowers *Rudbeckia* spp.
Butterfly Stonecrop (or Ice Plant) *Hylotelephium* [=*Sedum*] *spectabile*

Further reading and useful addresses

Identification of the more challenging genera and species is beyond the scope of this book and requires the use of a full key to all species with supporting descriptions and illustrations. For the British fauna, this is provided by Stubbs & Falk (2002), but other European works may also be useful. van Veen (2004) provides a key to the Western European species. This is somewhat more technical than Stubbs & Falk, but can be helpful if you get stuck! It also covers additional European species that could conceivably be found in Britain – so it is worth consulting if you think you have found something not currently covered by British texts. In the same vein, Martin Speight's project: 'Syrph the Net' seeks to covers the entire European fauna and now includes many keys as well as descriptions of the species and a wealth of other information. This is published as a series of spreadsheets and document files that are distributed electronically. See **http://www.iol.ie/~millweb/syrph/syrphid.htm** for contact information to request a copy. For the identification of larvae, Rotheray (1993) is really the only source.

The atlas (Ball, *et al.* 2011) provides distribution maps and information about status and trends in Britain, but maps quickly become out of date as new records arrive. The maps on the Hoverfly Recording Scheme web-site at **www.hoverfly.org.uk** should be checked for an up to date picture. The recent book by Rotheray & Gilbert (2011) provides a great deal of information about the biology and ecology of the family and, as mentioned above, Syrph the Net is another useful source for this sort of information, including the wider, European distribution and status of species.

For more general information about finding and catching flies, dealing with specimens, *etc.*, the 'Dipterists Handbook' (Chandler, 2010) is an invaluable source.

BALL, S.G. & MORRIS, R.K.A., 2000.
Provisional atlas of British hoverflies. Centre for Ecology & Hydrology, Huntingdon. 167pp.
This atlas is now largely out of date and more detailed maps can be found on the Hoverfly Recording Scheme website. Details of phenology remain useful.

BALL, S.G. & MORRIS, R.K.A., 2004.
A mark-release-recapture study of *Volucella bombylans* (Linnaeus, 1758), *V. inflata* (Fabricius, 1794) and *V. pellucens* (Linnaeus, 1758) (Diptera: Syrphidae). UK. *Dipterists Digest* (Second Series). **10(2)**: 73–83
This account describes techniques used to investigate hoverfly population biology. It is also useful because it illustrates how big hoverfly populations can be when their habitat appears to be scarce.

BALL, S.G. & MORRIS, R.K.A., 2006.
Britain's biggest hoverflies: the genus *Volucella*. *British Wildlife* **17**: 249–256.
Describes all six of the European species of *Volucella* and provides additional analysis of their ecology.

BALL, S.G., MORRIS, R.K.A, ROTHERAY, G.E & WATT, K.R., 2011. *Atlas of the Hoverflies of Great Britain (Diptera, Syrphidae)*. Biological Records Centre, Wallingford.

BALL, S.G. & MORRIS, R.K.A., (2014)
A review of the scarce and threatened flies of Great Britain:. Part 6: Syrphidae. *Species Status* **No. 9**: 1–124, Joint Nature Conservation Committee, Peterborough.
A detailed analysis of the trends within the British hoverfly fauna combined with allocation of statuses in accordance with the latest IUCN guidance. Available as a PDF from **www.jncc.gov.uk**.

CHANDLER, P. (ed.), 2010. A Dipterists Handbook (2nd ed.). *The Amateur Entomologist*, Vol. **15**. Amateur Entomological Society, Orpington. 525pp.
A comprehensive modern introduction to Diptera and their study.

DRAKE, M. & BALDOCK, N., 2005.
The bog hoverfly on Dartmoor. *British Wildlife* **17**: 102–106.

GILBERT, F.S. & FALK, S.J., 1986. *Hoverflies*. Naturalists' Handbooks 5.
Cambridge University Press. 72pp.
Provides keys to selected species. This is an excellent guide to the study of hoverflies.

LEVY, E.T., LEVY, D.A. & DEAN, W.F., 1989. *Dorset Hoverflies*. The Dorset Environmental Records Centre, Dorchester. 73pp.

LEVY, E.T. & LEVY, D.A., 1998. *Somerset Hoverflies*. E.T. & D.A. Levy, Yeovil.

MORRIS, R.K.A., 1998. *Hoverflies of Surrey*. Surrey Wildlife Trust, Pirbright. 244pp.
This is still the most comprehensive account of the fauna of a single county.

OWEN, J., 2010. *Wildlife of a Garden: A Thirty-year Study*. Royal Horticultural Society. 261pp.

ROTHERAY, G.E., 1993. Colour Guide to Hoverfly Larvae in Britain and Europe.
Dipterists Digest **9**. Derek Whiteley, Sheffield. (out of print).
This is a must have for the hoverfly enthusiast who wants to find and breed out hoverfly larvae.
www.dipteristsforum.org.uk/documents/DD/df_1_9_Colour_Guide_to%20Hoverfly_Larvae.pdf

ROTHERAY, G.E. & GILBERT, F., 2011. The Natural History of Hoverflies. Forrest Text, Cardigan.

STUBBS, A.E., 1991. A method of monitoring garden hoverflies. *Dipterists Digest* (first series) **10**: 26–39.
Still the only established technique for long-term surveillance monitoring of hoverflies.

SPEIGHT, M.C.D., 2011. Species accounts of European Syrphidae (Diptera), Glasgow 2011. Syrph the Net, the database of European Syrphidae, vol. 65, 285 pp., Syrph the Net publications, Dublin.

STUBBS A.E. & FALK, S.J., 2002. *British Hoverflies: an illustrated identification guide*. British Entomological & Natural History Society. 469pp. (Revised & Updated by Ball, S.G., Stubbs, A.E., McLean, I.F.G., Morris, R.K.A., and Falk, S.J.)
This is the essential companion to this **WILD***Guides* publication. Keys to all British species are accompanied by relevant sketches of important features. It is not possible to record the complete British hoverfly fauna without using this as the identification guide.

VEEN, M.P. VAN, 2004. *Hoverflies of Northwest Europe*. KNNV Publishing, the Netherlands.
Provides keys to many of the species that may ultimately occur in Britain. This is an excellent book, but not one for the novice as it is not comprehensively illustrated.

JOURNALS

DIPTERISTS DIGEST
published twice-yearly by Dipterists Forum. It frequently includes accounts of hoverfly ecology.
www.dipteristsforum.org.uk

HOVERFLY RECORDING SCHEME NEWSLETTER
published by the Recording Scheme as part of the Bulletin of the Dipterists Forum. Back numbers can be accessed from the Hoverfly Recording Scheme website.
www.hoverfly.org.uk/portal.php

BRITISH JOURNAL OF ENTOMOLOGY & NATURAL HISTORY
published by the British Entomological and Natural History Society. Four issues per year. Occasionally has articles on hoverflies.
www.benhs.org.uk/portal

BRITISH WILDLIFE
independent wildlife magazine, published bi-monthly. Includes regular reports on hoverflies and other flies. Available from: British Wildlife Publishing, Kemp House, Chawley Park, Cumnor Hill, Oxford OX2 9PH.
www.britishwildlife.com

USEFUL ADDRESSES

Dipterists Forum
This is the national society devoted to the study of flies. Its membership exceeds 400 people, most of whom take a non-vocational interest in Diptera. The Forum is extremely active, running a variety of field meetings and training events Novices are encouraged to participate in field meetings where it is possible to spend time with more experienced members and learn some of the fieldcraft needed to find hoverflies and other flies. The main Summer field meeting is usually based at a University or a field centre and runs for a week.
www.dipteristsforum.org.uk

UK Hoverflies Facebook page
This is a very active group with several members available to provide detailed advice on the identification of photographs. (Available to Facebook users only.)
www.facebook.com/groups/609272232450940/

Hoverfly Recording Scheme
Provides access to Recording Scheme outputs, news and announcements; provides discussion forums.
www.hoverfly.org.uk

Yahoo Hoverfly Group (UK Hoverflies)
Discussion forum for people interested in hoverflies.
groups.yahoo.com/neo/groups/UK-Hoverflies/info

The Malloch Society
This is a strictly Scottish society that was established to promote and undertake the study of flies. The society works as a team to undertake ecological studies of diptera in the UK, Europe and further afield.
www.mallochsociety.org.uk

Wild About Britain
Discussion forum for people interested in wildlife. Plenty of discussion about hoverflies.
www.wildaboutbritain.co.uk

iSpot
This website is dedicated to helping interested naturalists identify photographic records. The localities of the sightings are recorded so that data can be forwarded to or collected by Recording Schemes.
www.ispot.org.uk

Steven Falk's Hoverfly Photographs
This is a large resource of reliably identified photographs of hoverflies with species accounts.
www.flickr.com/photos/63075200@N07/collections/72157629600153789/

Acknowledgements

Our interest in hoverflies was largely influenced by Alan Stubbs' *British Hoverflies* and we are delighted to have Alan's foreword to this book. We must also thank Alan for encouraging us to take on the Hoverfly Recording Scheme back in 1991. We have gained immeasurable pleasure from such close involvement in hoverflies and we hope that this book does the same for a wide readership.

Many of the sections of this book have been tested on the many people who have attended the hoverfly identification courses that we run. Those that have attended our courses will doubtless recognise some of the descriptions. Our courses operate under the aegis of Dipterists Forum and there are numerous Forum members who have lent support to this work; indeed several very active Forum members joined because of these courses. This is greatly encouraging because we want to share our passion for hoverflies and to encourage a new generation of enthusiasts. Invertebrate conservation is greatly dependent upon maintaining detailed knowledge of the distribution and abundance of individual species.

The Hoverfly Recording Scheme can only function if there are enthusiasts who take the time to report their records. Over 3,000 individual recorders have contributed at least one record to the Scheme, including more than 1,700 since the year 2000. The maps used to illustrate this book derive directly from these data and we are immensely grateful to everyone who has contributed.

We are also indebted to various people who have drawn our attention to the need for corrections in the text of the first edition. In particular we thank Peter Chandler, Steve Franks and Ian Dawson who compiled comprehensive lists of errata. Development of this second edition has drawn upon the observations and comments of a number of contributors to the UK Hoverflies Facebook group: we are indebted to everybody who has engaged in this way.

The attractive presentation of the first edition received a good deal of positive feedback. This was substantially due to Rob Still's innovative layout and we are very grateful to him for the tremendous contribution he has made.

We would also like to thank Andy and Gill Swash, Brian Clews, Steve Holmes and John O'Sullivan for their help during the production process.

Photographic credits

The production of this book would not have been possible without the help and co-operation of a large number of photographers. For each of the species that we have treated in detail, we did our best to find one or more field photographs that illustrates its identification features. These are backed up by close-up shots of specimens to illustrate these features in detail. The specimens came mostly from the personal collection of Roger Morris, and their selection, cleaning and posing proved surprisingly time consuming! The photographs of specimens were taken by Stuart Ball and, for readers interested in macro photography, equipment from Canon was used: an EOS 60D camera, MP65 macro lens and MT-24EX twin flash unit. They are focus-stacked, were acquired using a StackShot electronically controlled focussing rail and processed using Helicon Focus and Zerene Stacker software.

The great majority of the field photographs were taken by Steven Falk. Whilst we spent innumerable hours searching the Internet for photographs, time and again we returned to Steve's images. The reason is clear: Steve is a hoverfly expert and his knowledge of the subjects enables him to capture the right angle and pose to show the identification features. His photos were therefore perfect for our purposes. We are also particularly indebted to Dr Brian Valentine who filled many of the gaps with a superb range of photographs. However, the works of 57 other photographers were essential to achieve coverage of the full range of species.

Photos of larvae are harder to find, so we are grateful to Dr Graham Rotheray for allowing us to use photographs from his *Colour Guide to Hoverfly Larvae*. Again, Brian Valentine's work was also very useful here and many of the field photographs of larvae are his.

The considerable time we have spent searching for photographs online has convinced us of two things: there are a lot of excellent pictures of hoverflies out there; and a significant proportion of them are incorrectly identified! It is a salutary experience to search Flickr, for example, for '*Eristalis tenax*' and see how many pictures of honey-bee are returned (or conversely, search for 'honey-bee' and see how many *Eristalis* come up!). We have made our own assessments and identifications of the photos we have chosen to use and take responsibility for the identifications given here.

Every photograph published in this book is credited in this section using the photographer's initials as follows: Stuart Ball [SGB], Tristan Bantock [TB], Ramon M Batlle [RMB], Ian Beddison [IB], Colin Boyd [CB], Graham Calow [GC], Ashley Cox [AC], Simon Damant [SD], Jelle Devalez [JD], Jeremy Early [JE], David Element [DE], Han Endt [HE], Steven Falk [SJF], David Fotheringham [DF], Penny Frith [PF], Peter Greenwoods [PG], Will George [WG], Elizabeth Grimes [EG], Andrew Halstead/Royal Horticultural Society [RHS], Håkon Haraldseide [HH], Louise Hislop [LH], Göran Holmström [GH], Nigel Jones [NJ], Maria Justamond [MJ], Bob Kemp [BK], Roger Key [RK], Stephen King [SK], Sergei Kuznetsov [SK], Jerry Lanfear [JL], Iain Lawrie [IL], Laurence Livermore [LL], Owen Llewellyn [OL], Richard Lyskowski [RL], Jess Maslen [JMa], Tony Matthews [TM], Ian McLean [IM], Jonathan Michaelson [JMi], Nigel Milbourne [NM], Roger Morris [RKM], John O'Sullivan [JO], John Pitts [JP], Adrian Plant [AP], Stephen Plummer [SP], Sandy Rae [SR], Tim Ransom [TR], Ben Revell [BR], Jeremy Richardson [JR], Ellen Rotheray [ER], Graham Rotheray [GR], Anne Sorbes [AS], Robert Still [RS], Malcolm Storey [MS], Andy and Gill Swash [A&GS], Mick Talbot [MT], Bill Urwin [BU], Brian Valentine [BV], Jaco Visser [JV], Walwyn [Wl], Ashley Wood [AW], Tim Worfolk [TW] and Cor Zonneveld [CZ].

Cover: *Episyrphus balteatus* – Quartl (licensed under the Creative Commons Attribution-Share Alike 3.0 Unported license and downloaded from http://commons.wikimedia.org/wiki/File:Episyrphus_balteatus_qtl2.jpg)

Frontispiece *Myathropa florea*: RS.

6 *Syrphus ribesii*: fem. on ragwort WG.
8 *Scaeva pyrastri* A&GS.
10 Marmalade fly *Episyrphus balteatus*: SGB.
12 *Syritta pipiens* TW.
13 *Syrphus* wing SGB.
14 Aphid feeding larva; Hoverfly ovipositing; Hoverfly puparium: *Syrphus* egg all BV; *Sphaerophoria* mating NJ.
15 *Volucella pellucens*; *Rhingia campestris* both BV.
16 *Eristalis nemorum* threesome JP.
17 *Melanostoma* killed by *Entomophthora* MS; Hoverfly eggs BV; Egg chorion SK.
18 *Epistrophe* larva BV.
19 Larva: feeding on aphids BV; *Dasysyrphus tricinctus* SGB; *Dasysyrphus venustus*; *Melangyna trianguliferum* both GR.
20 *Heringia* larva in *Pemphigus* gall; *Pemphigus* gall on poplar both SGB; *Microdon analis* larva in ant nest RL.
21 *Volucella inanis* larva RK.
22 *Portevinia maculata*: larva GR; adult SGB. Ramsons: SGB.
23 Finding larvae in thistle stems; *Cheilosia* larva *in situ*; *Cheilosia albipila*: larva; detail; *Cheilosia grossa* larva; detail – all images SGB.
24 *Cheilosia semifasciata*: leaf mine SGB; adult NJ. Wall Pennywort: SGB.
25 *Ferdinandea* larvae in sap run NJ; *Criorhina floccosa* larva; thorax of larva GR.
26 *Myathropa florea* larva GR. Rot hole in a Lime SGB.
27 Aspen log ER; *Hammerschmidtia ferruginea*: adult ER; larva GR. Garden pond SGB; *Eristalis tenax* larva GR.
29 *Scaeva pyrastri* BV.
30 *Myathropa florea* – both SGB.
31 *Merodon equestris*; f. *validus* NJ; f. *narcissi* (top right) NJ; f. *transversalis* SJF; f. *narcissi* (right) SJF; f. *equestris* SGB; f. *bulborum* SGB.
32 *Scathophaga stercoraria* eating hoverfly MJ; *Ectemnius* carrying hoverfly JE.
33 *Eristalis tenax*; Honey-bee – both SGB. *Volucella bombylans* NJ; Bumblebee SGB. *Chrysotoxum cautum* NJ; Wasp SGB.
35 *Cheilosia bergenstammi* SGB; *Eristalis intricaria*; *Dasysyrphus venustus* – both RKM.
36 *Brachyopa* tree RKM; *Epistrophe eligans* SGB.
37 *Criorhina floccosa* RK.
38 *Syrphus* on Angelica RKM.
39 *Myathropa florea* SGB.
40 Beech stumps SGB.
41 Old Beech SGB.
42 Chobham Common SGB; *Paragus* SJF.
43 *Sericomyia silentis*; *Lejops vittatus* – both SJF; Seasalter Marshes RKM.
44 *Platycheirus podagratus* SJF.
45 *Callicera rufa*; Artificial rot hole – both SGB.
49 *Cheilosia illustrata* SGB.
50 *Syrphus ribesii* – both SGB.
51 *Syrphus ribesii*; *Cheilosia* male head – all SGB.
52 *Syrphus* wing SGB.
53 *Syrphus ribesi*; *Syrphus* abd. – both SGB.
54 *Melanostoma scalare* fem. frons SGB. *Platycheirus albimanus* abd. dusting (Scanning electron micrographs) AP.
55 *Xylota florum*; *Syrphus ribesii* – both SGB.
56 Wings: *Psilota anthracina* SJF; *Eristalis horticola*; *Syrphus* (×2) – both SGB.
57 Wings: *Volucella bombylans*; *Neoascia tenur*; *Sphegina clunipes*; *Syrphus*; *Xylota* – all SGB. *Ferdinandea cuprea* fem.; *Xylota sylvarum* fem. – both SJF.
58 *Volucella bombylans* (×2) NJ; *Microdon analis* (×2) JV; *Eumerus* (×2) BV. *Volucella inanis* SJF.
59 *Callicera aurata*; adult; antennae – both SJF; *Sericomyia silentis* antennae SGB; *Cheilosia fraterna* antennae SJF; *Sericomyia superbiens* NJ; *Sericomyia silentis* SJF.
60 *Psilota anthracina* head SJF; *Pipiza luteitarsis* head;. *Rhingia campestris* head; *Pelecocera tricincta* head; *Portevinia maculata* head – all SGB.
61 *Cheilosia bergenstammi*; *Lejogaster metallina* – both SJF; *Cheilosia pagana* male face; *Cheilosia vulpina* head; *Lejogaster metallina* head – all SGB; *Brachyopa scutellaris* NJ.
62 *Platycheirus albimanus*; *Melanostoma mellinum*; *Meliscaeva cinctella*; *Paragus haemorrhous* – all SJF; *Syrphus ribesii* SK.
64 Solitary bees: *Andrena carantonica* TB; *Osmia rufa* BV.
66 Hornet *Vespa crabro* NJ; Common Wasp *Vespula vulgaris* SGB.
67–70 The most frequently photographed hoverflies – all SGB.
72 *Baccha elongata*; *Xanthandrus comtus* – both SJF; *Melanostoma mellinum* fem. frons SGB; *Platycheirus peltatus* male frons SJF.
73 *Melanostoma scalare* fem. abdominal markings; *Platycheirus albimanus* male; *Melanostoma scalare* male – all SJF; *Platycheirus albimanus* front leg SGB.

THE SPECIES ACCOUNTS

74 *Baccha elongata*: in flight OL.
75 *Baccha elongata*: male SJF; fem. BV.
77 *Melanostoma mellinum*: male SJF; male abd.; fem. fron SGB. *Melanostoma scalare*: male; fem. – SJF; male abd.; fem. frons – SGB.
78 *Platycheirus scutatus*: front leg SGB. *Platycheirus tarsalis*: male SJF.
81 *Platycheirus albimanus*: male SJF; front leg SGB; fem. (flying) SR. *Platycheirus ambiguus*: male SJF; front femur SGB.
83 *Platycheirus manicatus*: male SJF; front leg SGB. *Platycheirus tarsalis*: male SJF; front leg SGB.
85 *Platycheirus peltatus*: male; fem. SJF; front leg SGB. *Platycheirus scutatus*: male SJF; front leg SGB.
87 *Platycheirus angustatus*: fem. SJF; front tarsi SGB. *Platycheirus clypeatus*: male; fem. – SJF; front tarsi SGB.
89 *Platycheirus fulviventris*: male; fem. – SJF. *Platycheirus granditarsus*: male; fem. – SJF; male abd. SGB.
91 *Platycheirus rosarum*: male; fem. – SJF. *Xanthandrus comtus*: male SJF; fem. – JR.
92/3 *Paragus haemorrhous*: head SGB; male SJF.

94 *Xanthogramma pedissequum* NJ; *Syrphus ribesii* side view SGB; *Chrysotoxum arcuatum*; *Sphaerophoria scripta* – both SJF.
95 *Eriozona syrphoides* SR; *Leucozona glaucia* SJF; *Dasysyrphus albostriatus* IB; *Didea fasciata* wing; *Syrphus* wing – SGB.
97 *Parasyrphus punctulatus*: face TR; *Epistrophe diaphana* face SJF; *Episyrphus balteatus* abd. BV; *Eupeodes luniger* abd. JMi; *Melangyna umbellatarum* abd. SJF; *Scaeva pyrastri* abd. BV.
99 *Chrysotoxum bicinctum*: NJ.
Chrysotoxum festivum: EG.
101 *Chrysotoxum arcuatum*: SJF; antennae SGB.
Chrysotoxum cautum: CB;
antennae; male genital capsule – SGB.
103 *Chrysotoxum elegans*: JL; abd. SGB.
Chrysotoxum verralli: GC: abd. SGB.
105 *Xanthogramma citrofasciatum*: SJF.
Xanthogramma pedissequum: NJ.
106 *Conops quadrifasciata* (Conopidae): RK.
107 *Doros profuges*: DE.
109 *Sphaerophoria interrupta*: SJF;
female; side view; genitalia – SGB.
111 *Sphaerophoria rueppellii*: LL.
Sphaerophoria scripta: AC; spotted ind. TM.
113 *Eriozona syrphoides*: fem. SR.
Leucozona lucorum: male; fem. – SJF.
115 *Leucozona glaucia*: male IB; fem. AW.
Leucozona laternaria: SJF.
117 *Dasysyrphus tricinctus*: JL; wing SGB.
Dasysyrphus pinastri: head SGB.
Dasysyrphus albostriatus: IB.
118 *Dasysyrphus pinastri*: fem. CZ.
119 *Dasysyrphus venustus*: male; fem. SJF.
120 *Didea fasciata* wing SGB.
121 *Didea fasciata*: SR; halteres SGB.
Didea intermedia: HE; halteres SGB.
123 *Scaeva selenitica*: SR.
Scaeva pyrastri: BV; head SGB.
124 *Eupeodes corollae*: male abd.; head – SGB.
125 *Eupeodes luniger*: fem. SJF; frons SGB.
Eupeodes latifasciatus: male SJF; frons SGB.
Eupeodes corollae: frons SGB.
127 *Eupeodes corollae*: male; fem. SJF.
Eupeodes latifasciatus: male SJF; fem. SR.
128 *Melangyna* scutella – both SGB.
129 *Melangyna labiatarum*: fem. SJF; thorax SGB.
Melangyna umbellatarum: thorax SGB.
131 *Melangyna lasiophthalma*: SJF.
Melangyna umbellatarum: SJF.
133 *Melangyna cincta*: IB.
Melangyna quadrimaculata: male SR; fem. SJF.
135 *Meligramma guttatum*: DF; face DF.
Meligramma trianguliferum: WG.
137 *Meliscaeva auricollis*: fem.SJF; dark fem. BV; face SGB.
Meliscaeva cinctella: SJF; face SGB.
138 *Episyrphus balteatus*: BV.
139 *Episyrphus balteatus*: – both BV.
141 *Epistrophe nitidicollis*: SJF; scutellum SGB.
Epistrophe melanostoma: scutellum SGB.
Episyrphus squama; *Syrphus* squama – SGB.
143 *Epistrophe diaphana*: SJF; abd. SGB.
Epistrophe grossulariae: NJ; abd. SGB.
145 *Epistrophe eligans*: male; fem. SGB.
Megasyrphus erraticus: male TR.
147 *Parasyrphus punctulatus*: male; fem. – SJF; abd. SGB.
Parasyrphus vittiger: abd. SGB.
149 *Parasyrphus nigritarsis*: RMB; leg SGB.
Parasyrphus vittiger: SR; leg SGB.
Parasyrphus annulatus: leg SGB.
Parasyrphus malinellus: leg SGB.
151 *Syrphus ribesii*: SK; male leg; fem. leg – SGB.
Syrphus heads: male; fem. SGB.
Syrphus vitripennis: male leg; fem. leg – SGB.
152 2nd basal cells: *Syrphus torvus*;
Syrphus vitripennis – both SGB.
153 *Syrphus torvus*: male SJF.
Syrphus vitripennis: fem. SJF.
155 *Callicera rufa*: ER; Larva TR.
157 *Callicera aurata*: SJF.
Callicera spinolae: JO; inset SD.
158 *Portevinia maculata*; *Ferdinandea cuprea*; fem.; *Rhingia campestris* – SJF.
159 *Cheilosia chrysocoma* NJ; *Cheilosia bergenstammi* SJF; *Cheilosia vulpina* head SGB; *Portevinia maculata* head SGB;
Rhingia rostrata head SJF.
160 *Cheilosia pagana* male face SGB.
161 *Cheilosia illustrata* SJF; *Cheilosia scutellata* head HH; *Cheilosia albitarsis* legs BR; *Cheilosia proxima* head SJF; *Cheilosia vulpina* head SGB.
163 *Cheilosia caerulescens*: BV; Larval mines Andrew Halstead/RHS.
Cheilosia illustrata: SJF.
165 *Cheilosia variabilis*: SJF; face SGB.
Cheilosia vulpina: SJF; faces – both SGB.
167 *Cheilosia pagana*: MT; faces – both SGB.
Cheilosia longula: head SGB.
Cheilosia scutellata: HH; head SGB.
169 *Cheilosia albipila*: both images – SJF.
Cheilosia grossa: male SJF; fem. IL.
171 *Cheilosia bergenstammi*: SJF; head SGB.
Cheilosia fraterna: SJF.
173 *Cheilosia impressa*: male; fem. SJF.
Cheilosia proxima: male; fem. – SJF;
dusted sternites SGB.
175 *Cheilosia chrysocoma*: NJ.
Cheilosia albitarsis: BR; front tarsus SGB.
Cheilosia ranunculi: front tarsus SGB.
Tawny mining bee *Andrena fulva*: BV.
177 *Ferdinandea cuprea*: SJF.
Portevinia maculata: head SGB; male SP.
179 *Rhingia campestris*: SJF.
Rhingia rostrata: SJF.
180 *Sphegina clunipes* SGB; *Lejogaster metallina* abd. SGB; *Riponnensia splendens* abd. SJF.
181 *Riponnensia splendens*; *Chrysogaster solstitialis* male; *Melanogaster hirtella* head; *Chrysogaster cemiteriorum* head – SJF; *Myolepta dubia* abd.; *Brachyopa insensilis* – SGB.
183 *Neoascia meticulosa*: fem.; hind legs – SGB.
Neoascia podagrica: male SR; fem. SJF; wings SGB.
Neoascia tenur: hind legs SGB.
185 *Sphegina clunipes*: CZ; base of abd.; genitalia SGB.
Sphegina elegans: detail of thorax SJF.
Sphegina sibirica: SGB; base of abd. SGB.
Sphegina verecunda: genitalia SGB.
187 *Melanogaster hirtella*: male; fem. – SJF; face SGB.
Chrysogaster cemeteriorum: fem. abd. SGB.
Riponnensia splendens: fem. abd. SGB.
Lejogaster metallina: fem. abd. SGB.

189 *Chrysogaster cemiteriorum*: fem. SJF; thoracic pleura SGB.
Chrysogaster solstitialis: male; fem. SJF.
191 *Orthonevra nobilis*: SJF; head SGB.
Riponnensia splendens: male SJF; fem. LH; faces – both SGB.
193 *Lejogaster metallina*: fem. SJF; head SGB.
Lejogaster tarsata: CZ; heads – both SGB.
Soldier fly *Chloromyia formosa*: SJF.
195 *Brachyopa insensilis*: adult; head – SGB.
Brachyopa scutellaris: adult NJ; antenna SGB.
Brachyopa pilosa: antenna SGB.
Horse Chestnut sap run SGB.
196 *Hammerschmidtia ferruginea*: fem. SGB; male ER.
197 *Myolepta dubia*: SJF; abd. SGB.
198 *Merodon equestris* NJ; *Eristalis intricaria* SJF. *Eristalis tenax* SGB.
199 *Eristalinus sepulchralis* SJF; *Myathropa florea* thorax BV; *Helophilus pendulus* fem.; *Helophilus trivittatus*; *Anasimyia lineata* – SJF; *Parhelophilus frutetorum* male hind leg; *Anasimyia contracta* hind leg – SGB.
201 *Eristalis pertinax*: male SJF.
Eristalis horticola: wing; *Eristalis* faces – all SGB.
203 *Eristalis arbustorum*: male SJF;inset JMa.
Eristalis nemorum: SJF; pair SGB.
205 *Eristalis horticola*: SJF.
Eristalis rupium: TR.
Eristalis hind metatarsi – all SGB.
207 *Eristalis intricaria*: male; fem. SJF.
Eristalis tenax: SJF; face SGB.
209 *Eristalinus aeneus*: SJF; head BV.
Eristalinus sepulchralis: SJF.
211 *Mallota cimbiciformis*: SJF.
Myathropa florea: male SJF; fem.; wing SJB.
213 *Anasimyia lineata*: SJF; head SGB.
Anasimyia contracta: male; fem. – SJF.
215 *Helophilus hybridus*: all images SGB.
Helophilus pendulus: male; fem. SJF; fem. frons; fem. hind leg – SGB.
217 *Helophilus trivittatus*: SJF.
Helophilus pendulus: fem. face SGB.
Lejops vittatus: SJF.
219 *Parhelophilus frutetorum*: SJF; hind leg SGB.
221 *Eumerus ornatus*: male. SJF; fem. head SGB.
Eumerus funeralis: male BV; fem. JR; hind leg; f em. head – SGB.
Eumerus strigatus: hind leg SGB.
223 *Merodon equestris*: tawny morph SJF; hind leg SGB.
Eumerus sabulonum: TB.
224 *Hydrotaea aenescens* (Muscidae): HH.
225 *Psilota anthracina*: SJF.
227 *Pelecocera tricincta*: SJF; head SGB.
Pelecocera scaevoides: IM; on grass TR.
228 *Pipiza luteitarsis* head SGB; *Pipiza austriaca*; *Psilota anthracina* (×2) – SJF.
229 *Triglyphus primus* abd.; *Pipizella viduata* abd.; wing; *Pipiza noctiluca* wing – SGB.
231 *Pipiza austriaca*: SJF; hind leg SGB.
Pipiza luteitarsis: SJF; head SGB.
233 *Pipiza noctiluca*: SJF.
235 *Heringia* profile; *Pipiza luteitarsis* profile – SGB; *Heringia heringi*: male SJF; *Heringia* (*Neocnemodon*) sp. SJF; trochanters SGB.
236 Solitary bee *Lasioglossum* sp.: BV.
237 *Pipizella viduata*: SJF; head profile; genitalia – SGB.
Pipizella virens: head profile; genitalia SGB.
239 *Trichopsomyia flavitarsis*: both – SJF.
Triglyphus primus: SJF.
240 *Sericomyia superbiens*: head SGB.
241 *Sericomyia superbiens*: NJ.
243 *Sericomyia lappona*: CB.
Sericomyia silentis: SJF.
244 *Volucella pellucens* head; wing – SGB.
245 *Volucella bombylans*: NJ.
Volucella bombylans: var *plumata* SJF.
Bombus lapidarius: PG.
Bombus lucorum: agg. SGB.
247 *Volucella inflata*: SJF.
Volucella pellucens: male JMi; fem. SJF.
249 *Volucella inanis*: SJF; sternites SGB.
Volucella zonaria: SJF; sternites SGB.
250 *Caliprobola speciosa* AS; *Blera fallax* ER; *Syritta pipiens*; *Tropidia scita* – SJF; *Syritta pipiens* hind leg; *Tropidia scita* hind leg – SGB.
251 *Criorhina ranunculi*; *Pocota personata*; *Criorhina asilica*; *Brachypalpus laphriformis*; *Xylota segnis* – SJF; *Brachypalpoides lentus* BK.
252 *Xylota* wing SGB.
253 *Blera fallax*: male; fem. ER.
255 *Brachypalpoides lentus*: BK.
Caliprobola speciosa: BU.
Sawfly *Macrophya annulata*: Wl.
Spider-hunting wasp *Priocnemis perturbator*: JD.
Old beech Highstanding Hill Windsor SGB.
256 Solitary bee *Osmia* sp.: GH.
257 *Brachypalpus laphriformis*: SJF; hind leg; metasternum SGB;
Chalcosyrphus eunotus: both images SGB.
259 *Chalcosyrphus eunotus*: BK.
Chalcosyrphus nemorum: SJF; hind leg; metasternum; *Xylota abiens*: hind leg – SGB.
261 *Criorhina* head SGB.
Criorhina asilica: SJF.
Criorhina ranunculi: male IB; fem. SJF.
263 *Criorhina berberina*: Wl; form *oxyacanthae* SJF.
Criorhina floccosa: SJF.
Bombus pascuorum: PF.
264 *Pocota personata*: mating pair – both images JR.
265 *Pocota personata*: both images SJF.
267 *Syritta pipiens*: BV.
Tropidia scita: SJF.
269 *Xylota jakutorum*: SJF.
Abdomens; hind femurs – all SGB.
271 *Xylota segnis*: SJF.
Xylota sylvarum: SJF. Hind legs – all SGB.
273 *Microdon analis*: JV; puparium; puparia under bark – SGB.
275 *Microdon devius*: BK; thorax SGB.
Microdon myrmicae: SJF; thorax SGB.
276 *Chrysogaster solstitialis* SJF.
285 Using a hand lens SGB.
286 Roger Morris using long handled net SGB; Stuart Ball in Scotland RKM; pooter; detail of pooter bung – SGB.
287 pinned specimen; specimen labels – SGB.
289 store box SGB.
290 *Eristalis cryptarum* SGB.
292 *Episyrphus balteatus* on Busy Lizzy BV.
293 Mature larva of *Epistrophe* BV; *Myathropa* nest box SGB
294 Roger Morris SGB

Index of scientific names

This index includes the *scientific* names of all the hoverfly species covered in this book.

Bold black text is used for species that are afforded a full account;
bold brown is used for tribes and sub-families.

Bold black numbers indicate the main species account.

Italicized red numbers relate to other photographs of adult, or parts of adult, insects.

Italicized green numbers relate to photographs of early life stages (larvae or eggs).

Numbers in 'normal' text refer to other pages where the species is mentioned.